AF597675

HAZARDOUS WASTE MANAGEMENT

HAZARDOUS WASTE MANAGEMENT

Edited by
J. Jeffrey Peirce
P. Aarne Vesilind

230 Collingwood, P.O. Box 1425, Ann Arbor, Michigan 48106

Library of Congress Card Catalog Number 81-67509
ISBN 0-250-40459-1

Manufactured in the United States of America

Butterworths, Ltd., Borough Green, Sevenoaks, Kent TN15 8PH, England

PREFACE

This book is intended for all readers concerned with hazardous waste management strategies, but especially for engineers in consulting practice, working for industry, or staffing federal, state or local government agencies. While much of what we have written applies to those who must cope with the discovery of abandoned disposal sites, many of our chapters are applicable to more general hazardous waste problems. Current generators, transporters, processors and disposers of the waste will benefit from all sections of the book, as will elected officials and town managers faced with key issues raised by well informed constituencies.

The papers in this volume were originally prepared for a conference sponsored by the Engineering Foundation and held on the campus of Franklin Pierce College in Rindge, New Hampshire. The idea and inspiration for the conference came from Earl Jones, of the Department of Housing and Urban Development. The conference had three co-chairmen, with Clinton Parker of the University of Virginia joining the editors in that capacity.

The editors and authors of the individual chapters wish to thank the Engineering Foundation for the chance to exchange ideas at the conference. Each chapter was molded by formal and informal interactions between participants and those in attendance during that rainy week of August 18, 1980. For this, we also wish to thank the attendees, too numerous to list here; they added greatly to the success of the conference and the quality of this book.

J. Jeffrey Peirce
P. Aarne Vesilind

ACKNOWLEDGMENTS

The authors wish to thank the Engineering Foundation for the support and encouragement before, during and after their August 1980 conference entitled *Hazardous Waste Management: Old Problems and New Solutions*. This volume summarizes the discussions that took place at that conference, and is intended to extend the proceedings to engineers charged with hazardous waste management responsibilities who were unable to attend.

Additionally, the authors wish to thank those individuals who directed the discussions during the week-long session. Without their input and guidance, the conference and this book would not have been possible. Special thanks go to Jim Murphy for his view of hazardous waste as expressed in our epilogue.

Peirce

Vesilind

J. Jeffrey Peirce holds a Bachelor of Engineering Science in engineering mechanics from The Johns Hopkins University. He received his MSCE and PhD in civil and environmental engineering/economics at the University of Wisconsin, Madison. As a faculty member at Duke University, Dr. Peirce lectures on topics ranging from air pollution abatement technologies to resource recovery systems, and conducts research within the Department of Civil Engineering. Dr. Peirce's current research develops and compares hazardous waste disposal options for Duke University, analyzes data on municipal sludge to provide a basis for future disposal options, and studies less costly alternatives for regional solid waste disposal and resource recovery. He has more than 20 publications in such journals as *Water Resources Bulletin* and the *Journal of the Environmental Engineering Division of ASCE.* Dr. Peirce is a full member of Sigma Xi, the American Society for Engineering Education and the Association of Environmental Engineering Professors.

P. Aarne Vesilind is Associate Professor of Civil Engineering at Duke University, Durham, North Carolina, and Director of the Duke Environmental Center. Born in Estonia, he holds BS and MS degree in civil engineering from Lehigh University, and a Master's degree and a PhD in sanitary engineering from the University of North Carolina. Dr. Vesilind has also been a Fellow at the Norwegian Institute for Water Research in Oslo, and has received a Fulbright-Hayes Senior Lectureship to study in New Zealand.

Among his many honors and awards are the Duke University Outstanding Professor Award for 1971–1972 and the 1971 Collingwood Prize from the American Society of Civil Engineers.

Dr. Vesilind is the author of two Ann Arbor Science publications, *Environmental Pollution and Control* and *Treatment and Disposal of Wastewater Sludges*, is the series editor of the *Design and Management for Resource Recovery* series, and is a member of the Ann Arbor Science Publishers Editorial Advisory Board.

CONTENTS

Part 3
Case Studies in the Management of Hazardous Waste

INTRODUCTION

J. Jeffrey Peirce

Department of Civil Engineering
Duke University
Durham, North Carolina

Hazardous waste managers are confronted with a complex set of old problems and few new solutions. For centuries, chemical wastes have been the necessary by-product of developing societies. Here a disposal site; there a disposal site; everywhere a disposal site—all with little or no attention to potential impacts on public health and environmental quality. Management decisions historically were made by default; little or no disposal planning at the corporate or plant level necessitated quick and dirty decisions by middle management at the end of production processes. Production engineers solved disposal problems by simply piling or dumping the waste products "out back."

Attitudes began to change during the 1960s and 1970s. Air, water and land no longer were viewed as commodities that could be polluted and freely passed to neighboring towns or future generations: individuals responded with court actions against pollution; government responses ranged from revised local zoning ordinances to new state public health laws to major federal Clean Air and Clean Water Acts. In 1976 the federal Resource Conservation and Recovery Act (RCRA) was enacted to give the U.S. Environmental Protection Agency (EPA) specific authority to regulate the generation, transport and disposal of hazardous waste. As the

1980s begin, we find that engineering knowledge and expertise lag behind this new awakening for the necessity to adequately manage hazardous wastes disposal.

Engineering Foundation Conferences are organized to provide an opportunity for the exploration of such problems addressed by engineers from many disciplines. The conferences, held throughout the year but typically at a mountain retreat during the summer weeks, are designed to encourage discussions of recent developments in technology and provoke suggestions concerning innovative methods of approaching the topic of the week. Attendees at any one conference generally represent a cross section of the national societies that support the conferences: ASCE, AIChE, ASME, AIME and IEEE.

This book summarizes conference discussions that took place in Rindge, New Hampshire, during the week of August 17 through August 22, 1980, under the title: *Hazardous Waste Disposal: Old Problems and New Solutions*. The organization of the conference was based on the premise that:

- many abandoned disposal sites exist but have not yet been discovered;
- individuals and/or firms may or may not know where these sites are located;
- public health and environmental quality will be impacted;
- the abandoned sites will be discovered; and
- engineers from all sectors will be called on to respond:
    - Engineers in industry: What can we do?
    - Engineers in consulting practice: How can we help?
    - Engineers in government service: What must we do?

The conference was thus structured to assist engineers in all fields to answer these key questions.

Our presentation is organized into sections according to three major topics: (1) Problems in Hazardous Waste Management; (2) Technological Inputs to the Management Process; and (3) Case Studies in the Management of Hazardous Wastes. Below, we briefly introduce these major topics.

PROBLEMS IN HAZARDOUS WASTE MANAGEMENT

Problems in hazardous waste management, particularly problems associated with the discovery of an abandoned disposal site, cover the entire spectrum from "What is a hazardous waste" to "Who is going to pay for this clean-up." Each chapter under this broad heading is introduced below.

- *Hazardous Waste Identification.* Issues surrounding the classification and categorization of hazardous waste are addressed initially. Emphasis is placed on current philosophies as reflected in RCRA, as well as a consideration of the degree to which the RCRA system can and will work in practice.
- *Historical and Ethical Perspectives.* Secondly, we consider the responsibilities of hazardous waste managers, not in the legal and administrative sense, but from the ethical standpoint: Why should we care about hazardous waste and dirty water and dirty land anyway? We find it much easier, for example, to agree that Kepone® should be classed as a hazardous waste than to explain why we should be in the least concerned about the fathead minnow, the blue crab or the Canadian geese that inhabit the James River Estuary.
- *Public Health Impacts.* In this initial section, we also address the public health impacts of improper disposal of chemical wastes. We find that acute effects on health have been documented and, in many cases, the nature of the physiological problems is consistent with the nature of the known contaminants. We note that the long-term health problems have not been studied adequately, but effects most probably will involve a range of physiological systems—genetic, hepatic and respiratory. Psychological effects have not received attention until recently, but for large-scale incidents they appear serious and costly.
- *Decision-making in Hazardous Waste Disposal.* There are almost as many ways to make decisions as there are decision-makers. In this chapter, we present an analysis limited to rational procedures of decision-making: to the characteristics that make them rational, to a methodology for analyzing these procedures and determining the values contained within them, and to an assessment of whether the final decision is congruent with that set of determined values.

We end this section with suggestions on where the decision-maker can look for guidance, and we note that one critical input to the hazardous waste management process is the availability of technology. The second major topic of this book deals directly with technology considerations.

TECHNOLOGICAL INPUTS TO THE MANAGEMENT PROCESS

A wide range of technologies are available to hazardous waste managers as they address the cleanup of abandoned sites or the operation of ongoing processing/disposal facilities. This section of the book addresses four major considerations.

- *Hazardous Waste Incinerators.* The attractiveness of high-temperature incineration is increasing across the nation. Stringent landfill regulations and rising public objection even to "adequate" landfills have sharply reduced industry's ability to follow traditional methods of waste disposal. This chapter addresses three approaches to industrial waste incineration.

- *In Situ Treatment/Containment and Chemical Fixation.* In situ physical containment and in situ treatment are two methods that are available for use at hazardous waste dump or spill sites. We address these two categories of waste management technology and further elaborate on chemical fixation processes available for use on hazardous wastes.
- *Siting Waste Management Facilities.* Critical to any hazardous waste management scheme is the siting of management facilities. Our purpose in this chapter is to examine the siting issue in detail and to propose some solutions to the problems encountered across the nation. We focus on four categories of siting issues: technical, economic, institutional and political.
- *Superfund Legislation.* This chapter addresses the concept of a "superfund" to provide funds to control the releases of hazardous substances into the environment. The concept is presented in terms of how the fund would work, the size of the fund, what parties should contribute to the fund, third party damage claims against the fund, and legal liability provisions.

This second section attempts to address a range of technologies available to hazardous waste managers. Possibly of equal value to the manager as he addresses current problems is the Case Studies discussion included in the final section of this book.

CASE STUDIES IN THE MANAGEMENT OF HAZARDOUS WASTE

In this section, we present three case studies in the management of hazardous waste. A wide range of problems is considered from the Kepone cleanup in Virginia, to the Ohio River Park on Neville Island near Pittsburgh to a runaway chemical reaction started in a plant near Milano, Italy. Cases were selected for inclusion to array the problems encountered and possible solutions available as past hazardous waste management decisions have been made.

- *The Kepone Cleanup, Hopewell, Virginia.* On May 28, 1975, a worker at the Hopewell, Virginia, Sewage Treatment Plant was overcome by fumes from hexachlorocyclopentadiene (HCP). The incident was reported to the State Water Control Board and the State Health Department, and Kepone became a household word across the nation. This case study illustrates government, industry and public response to this serious problem.
- *Ohio River Park, Allegheny County, Pennsylvania.* In 1976 a 35-acre site at the western tip of Neville Island in the Ohio River, Pennsylvania was donated to Allegheny County for the purpose of developing a recreational park. The discovery of an abandoned hazardous waste disposal site at the location is documented, and the response is discussed in terms of available technologies and government/private reactions to the events that followed the discovery.

- *Experience of the Accident at Seveso.* On July 10, 1976, a runaway chemical reaction started in a plant that was synthesizing trichlorophenol. The safety valve of the reactor was not provided with protective devices, and hot vapors blew into the atmosphere; the resulting cloud, moved by a weak wind, began to fall to the ground. This case study addresses public health impacts of the accident.

In aggregate, the case studies display the range of difficulties that an engineer must confront as he seeks solutions to his particular problem(s). Much can be learned from these past responses to hazardous waste management problems.

PART 1

PROBLEMS IN HAZARDOUS WASTE MANAGEMENT

CHAPTER 1

CLASSIFICATION AND CATEGORIZATION OF HAZARDOUS WASTE

John H. Frick

Waste Disposal Engineering Division
U.S. Army Environmental Hygiene Agency
Aberdeen Proving Ground, Maryland

The Resource Conservation and Recovery Act (RCRA) of 1976 set up the basic structure of a program to regulate the handling and disposal of solid waste and to encourage and regulate resource recovery and reuse. Solid waste has been defined in the act to include both hazardous and nonhazardous wastes. However, because of the radically different environmental effects of these two classes of wastes, two markedly different statutory and regulatory controls were developed.

The U.S. Environmental Protection Agency (EPA) was assigned the tasks of defining the term "hazardous waste" and then developing a comprehensive program to minimize its adverse impacts on the environment. The program was designed to address all stages of hazardous waste management from generation to ultimate disposal, i.e., "cradle-to-grave." However, one of the most critical features of this regulation was to provide a suitable definition for hazardous waste.

The regulations proposed December 18, 1978 identified hazardous wastes by providing lists, test methods for hazard characteristics or by declaration. The listing method was one good way of identifying hazardous wastes because many types of wastes (e.g., industrial wastes) tend to be complex mixtures of numerous components; and by searching the lists, the waste generator could determine if his waste was hazardous without excessive testing of the hazard characteristics. Also, this method was simple and eliminated a few of the uncertainties in declaring a waste hazardous. The selection of items to be listed was based partly on incident reports and partly on generalized information that had revealed the dangerous characteristics and risks of a waste when improperly handled or managed. Numerous cases of faulty waste management resulting in extensive damage to public health and the environment have been documented (e.g., Love Canal and Valley of the Drums). It was these types of damage cases coupled with general reference material that EPA used when selecting items for placement on the lists.

The tests used to determine hazard characteristics (i.e., ignitability corrosivity, reactivity and toxicity) were selected from well-established standard methods (e.g., ASTM); but problems were associated with the use of simple test protocols for single characteristics in that they are usually too restrictive in scope and confine themselves to measuring only one condition or stress.

After the December 18, 1978 publication of the proposed rules, EPA received a large number of comments forcing the publication of the final rules two years behind the schedule spelled out in RCRA. Consolidated EPA efforts and a rush to meet court order deadlines resulted in publishing the final rules on May 19, 1980. According to some these final rules are the most complex set of regulations yet assembled by EPA. These rules govern more than 500 specific wastes, waste processes and waste sources and will regulate about 40 million metric tons of the estimated 57 million metric tons of hazardous waste generated nationwide each year.

As a result of the complexity and the pressures to expedite the publication of these rules, certain waste streams addressed in the proposed rules were not identified in the final rules (e.g., waste oils, PCB, infectious wastes, paint residues and low-level radioactive wastes). Currently, EPA is gathering more information on these waste streams in hopes to address them or aspects of them in the second phase publication of these regulations.

Because of increasing public concern over environmental problems related to hazardous waste disposal, citizens will continue to press for improved management practices and technology. It is anticipated that waste materials will be added to the hazardous waste listing throughout the 1980s. Petitions for changes in the regulations and delisting of facility waste as hazardous waste should be commonplace. The attack on hazardous waste is surely one of the major challenges of the 1980s.

OVERVIEW

Hazardous waste management is the preeminent environmental priority today and will be a major challenge for society to bring under control during the 1980s. It is clear that dealing with this challenge is everybody's problem. It is not just a government problem. It is not just an industry problem. It is a problem for all citizens in our society.

In the process of producing goods and services, we also produce wastes and, in many cases, the wastes we generate are hazardous. For example, the petrochemical industry produces such hazardous wastes as phenols, heavy metals, acids, caustics and organic compounds. In the production of goods involving metals, we produce such hazardous wastes as heavy metals, fluorides, cyanides, acids, alkaline cleaners, solvents and phenols. In the process of producing leather goods, we produce such hazardous wastes as heavy metals and sulfides.

Data presently available indicate that about 90% of the hazardous waste generated in the United States is managed in ways that are not totally sound and that only about 10% is managed within the safeguards embodied in the hazardous waste regulations. Much of our hazardous waste management, however, can be upgraded to meet new standards.

For years we did not worry about this. Environmentally, we focused on different problems such as dirty smoke and discharge to waterways. Both were highly visible and media that are common to us all. Waste was a neglected issue. Once in the ground it was "out-of-sight, out-of-mind." It was not an issue to worry about.

Elucidation of the "Love Canal" problem in August, 1978 changed it all and 230 families were evacuated by the end of that summer. Recently, 700 families were moved from the site. Clearly, this is the first in rank of environmental tragedies to come in our country. Another problem area was the Chemical Control Facility at Elizabeth, New Jersey. When the state of New Jersey took over, about 35,000 barrels of waste were stored there. Fortunately, the state was able to put together enough resources to remove some of the most hazardous materials before the facility blew up. Open dumping and mismanagement of hazardous wastes are, unfortunately, all too typical of practices that are customary.

As a result of not paying attention to this problem, what have we inherited? We have inherited thousands of orphan disposal sites and billions of dollars of debt for containment and cleanup. We know the number of sites and where wastes have been disposed of improperly. We know the estimated cost of cleanup. Also, we have seen the risks and damages to the public health (e.g., about 14 billion in law suits is pending from the Love Canal incident).

Why did this happen? A deficient management system equals a nonsystem. Except for a few states, the identification of waste is poor. There is inadequate assignment of responsibility for the generators, transporters and owners and operators of treatment, storage and disposal facilities. There are no standards for these facilities. There is inadequate information on the size, scope and nature of the problem. There is no notification system for emergencies such as road spills. Also, there are inadequate incentives to encourage best thinking and resource utilization to solve these problems.

As we move to the 1980s we face a double challenge: legacies of the past, a nonsystem; and a system for safe management of hazardous waste.

IDENTIFICATION

Hazardous waste is defined in the Resource Conservation and Recovery Act of 1976 and in Part 261 of the Act's Subtitle C regulations. Hazardous waste is identified by listing and testing characteristics. Subpart A of Part 261 identifies wastes that are excluded from these regulations and it establishes special management requirements for small generators and hazardous wastes that are used, reused, recycled or reclaimed. Subpart B of Part 261 sets forth the criteria that EPA will use to identify the characteristics of hazardous waste and to list certain solid wastes as hazardous wastes. Subpart C identifies the characteristics of hazardous waste and subpart D lists particular hazardous wastes. It is important to note that Part 261 identifies only some items that are hazardous wastes under Sections 3007 (Inspections) and 7003 (Imminent Hazard) of RCRA. A material not identified in this part is still a hazardous waste if EPA has reason to believe the material may be hazardous under Section 1004 (Definitions) of RCRA; or, if it is established under Section 7003 (Imminent Hazard) of RCRA.

A hazardous waste is defined as subset of all solid waste. A solid waste is any garbage, refuse, sludge or other waste material not excluded (i.e., domestic sewage, irrigation return flows, source, special nuclear or by-product radioactive wastes and onsite mining materials). Garbage, refuse, sludge and other waste materials are solid wastes, whether discarded, used, reused, recycled, reclaimed, stored or accumulated for any of these purposes.

What is other waste material? The Act indicates under an explanation of other discarded material that it includes solid, liquid, semisolid and contained gaseous material resulting from industrial, commercial, mining or agricultural operations, or from community activities. EPA further adds that other waste material is that which is discarded or intended to be discarded, has served its original intended purpose, or is a manufacturing or mining by-product. Discarded material is that which is abandoned (i.e., not used, reused, reclaimed or recycled) by being disposed of, burned or incinerated except for energy recovery, or treated by some other method before being discarded.

Specific materials that RCRA excluded as solid waste are: domestic sewage, irrigation return flows, industrial wastewater point source discharges, special nuclear and by-product radioactive materials, and onsite mining wastes that are not removed from the ground as part of the extraction process. It is important to note that the point source discharge exclusion applies to the actual point source discharge from an industrial waste treatment plant, not the collection, storage, treatment processes or the generated sludges. A solid waste is broadly defined in that it does not have to be a solid; only those items that are discarded.

The definition of hazardous waste includes the following:

1. It is first a solid waste.
2. It is not excluded from the regulation.
3. It is any of the following:
   (a) a listed hazardous waste,
   (b) a mixture containing a listed hazardous waste, or
   (c) an unlisted waste having any of the four identified characteristics (ignitability, corrosivity, reactivity or extraction procedure toxicity).

When does a solid waste become a hazardous waste? For listed waste, it becomes a solid waste as soon as it meets the definition of a listed waste. EPA has indicated its basis for listing classes or types of wastes. Appendix VII (of RCRA) identifies the constituent that caused EPA to list the waste as toxic waste (T). Other listed wastes are designated by specific tests (i.e., ignitability, corrosivity, reactivity or extraction procedure toxicity) or designated by the administrator as acute hazardous waste (H). The latter are commercial products subject to the small quantity exclusion (i.e., 1 kg, 20-liter containers, 10-kg liners and 100-kg residues). For a mixture, a waste will become a hazardous waste as soon as a listed waste is added to the mixture. For any other waste it will become hazardous when the waste exhibits a hazardous waste characteristic identified in subpart C, i.e., ignitability, corrosivity, reactivity or extraction procedure toxicity.

When does a hazardous waste cease being a hazardous waste? When it is delisted. Delisting can be accomplished only when one successfully petitions to exclude a waste at a particular facility. A petitioner must demonstrate that his waste does not meet any of the criteria under which the waste was listed. Therefore, a hazardous waste is hazardous as long as it is stored or disposed of, if removed from storage or disposed of, as long as it is being treated, and after treatment, unless delisted (e.g., waste battery acid that is accumulated, stored, neutralized and disposed of remains a hazardous waste even after neutralization treatment, unless properly delisted). Also, residues from hazardous waste treatment, storage or disposal facilties are hazardous unless delisted (e.g., sludges, treatment residues, spill residues, ash, air emission control sludge/dust and leachate).

Hazardous waste exclusions include the following:

1. household wastes (i.e., garbage, trash and sanitary wastes in septic tanks derived from residences, hotels and motels),
2. municipal resource recovery wastes, i.e., refuse-derived fuel (RDF);
3. agricultural residues (i.e., those returned to the soil as fertilizers); and
4. mining overburden returned to the mine site.

Temporary exclusions include the following:

1. utility wastes (i.e., fly ash, bottom ash waste, slag waste and flue gas emission control waste from fossil fuel combustion; and
2. drilling fluids, brines and other waste associated with oil and gas production, exploration and development and geothermal energy recovery.

Under subtitle C regulation, special requirements for hazardous waste generated by small generators do not apply if less than 1000 kg/month is generated or less than 1000 kg total is accumulated; or provided the waste is taken to a subtitle C treatment, storage or disposal facility or a facility that manages municipal or industrial solid waste and is permitted by the state. Special requirements also apply to generators of certain acutely hazardous waste. In cases of the latter, special requirements apply to 122 substances listed as "P" substances. These requirements do not apply to generators, transporters, and owners and operators of storage, treatment or disposal facilities or for permitting permission and notification requirements if the quantity of the waste item is less than the following:

- 1 kg/mo of the listed waste,
- 1 kg/mo of off-specification national,
- 20-liter container of the waste,
- 10 kg/mo of inner liners, or
- 100 kg/mo of spill residue.

In addition, special requirements are presented for hazardous wastes that are used, reused, recycled or reclaimed. These hazardous wastes are not now regulated except for sludges or listed hazardous wastes that are stored or transported prior to being used, reused, recycled or reclaimed. In the latter case, these items are fully regulated under subtitle C storage, transportation, notification and generation regulations.

CHARACTERISTICS

Part 261, subpart C discusses characteristics of hazardous waste and includes the criteria for identifying the characteristics of hazardous waste. A solid waste is identified as a hazardous waste if it:

1. meets the statutory definition of hazardous waste;
2. is measurable by standarized test methods and is available and within the capacity of generators or private sector laboratories; or
3. can be reasonably detected/determined based on knowledge about the waste.

The characteristics testing for determining a waste as hazardous includes the following: ignitability, corrosivity, reactivity, and extraction procedure toxicity. It should be noted, however, that a solid waste that exhibits any of these characteristics is a hazardous waste, whether it is listed or not.

Ignitability

A solid waste exhibits the characteristics of ignitability if a representative sample of the waste has the following properties:

1. a liquid with a flash point less than 60°C (140°F) as measured by the Pensky-Martens closed cup tested method or Setaflash closed cup test method (ASTM Standards), which applies to any liquid except aqueous solutions containing 24% alcohol by volume;
2. a nonliquid that is capable under normal conditions of spontaneous and sustained combustion;
3. ignitable compressed gas; and
4. oxidizer.

Waste meeting this characteristic is designated by an EPA hazard code "I" and is identified by the number D001.

Corrosivity

A solid waste exhibits the characteristic of corrosivity if a representative sample of the waste has either of the following properties:

1. liquid with a pH less than 2 or greater than 12.5; or
2. liquid that corrodes steel at a rate greater than 1/4 inch per year.

The method described in EPA 600/4-79-020 using a pH meter or the method described by the National Association of Corrosion Engineers (NACE), Standard TM-01-69, or equivalent methods can be used when approved by EPA. Waste meeting this characteristic is designated by an EPA Hazard Code "C" and is identified by the number D002.

Reactivity

A solid waste exhibits the characteristic of reactivity if a representative sample of the waste has any of the following properties:

1. normally is unstable—reacts violently;
2. reacts violently with water;
3. forms explosive mixtures with water;
4. when mixed with water generates toxic gases, vapors or fumes;
5. contains cyanide or sulfide and generates toxic gases, vapors or fumes between pH 2 and 12.5;
6. could detonate if heated under confinement or subject to strong initiating source;
7. could detonate at standard temperature and pressure; and
8. is listed by the Department of Transportation (DOT) as Class A or B explosive. Waste meeting this characteristic is designated by an EPA hazard code "R" and identified by the number D003.

The only testing associated with this hazard class is that required by DOT. EPA has used standardized testing protocols to numerically quantify its properties; however, available test methods have had many shortcomings. The tests are too restrictive. They confine themselves to measuring how one aspect of reactivity correlates to one condition or stress that initiates the reaction. Reactivity is a function of both density and composition but also mass and surface area. A single test will not measure the whole. Most tests are not "pass-fail" type but require subjective interpretations. Also, most reactive waste generators are aware of the properties of their wastes because of safety requirements. Consequently, a prose definition should allow generators to determine whether their wastes are reactive.

Extraction Procedure Toxicity

A solid waste exhibits characteristics of EP toxicity if, using the described extraction procedure or an equivalent method, the extract of a representative sample contains a maximum concentration as given below for the eight heavy metals and six pesticides.

Heavy Metals	mg/l
Arsenic	0.5
Barium	100.0
Cadmium	1.0
Chromium	5.0
Lead	5.0
Mercury	0.2
Selenium	1.0
Silver	5.0

Pesticides	**mg/l**
Endrin	0.02
Lindane	0.4
Methoxychlor	10.0
Toxaphene	0.5
2,4-D	10.0
2,4,5-T	1.0

The concentrations given for each item are 100 times the primary drinking water standards. Also, no extract is necessary for the extraction procedure toxicity test if the waste is a liquid with less than 0.5% filterable solids. In this case, the waste itself is considered the extract. Waste meeting this characteristic is assigned a hazard code "E" and identified by numbers D004-D017.

LISTS

A waste may be listed as a hazardous waste if it meets the statutory definition of hazardous waste, i.e., if it causes or contributes to:

1. increased mortality,
2. increased serious irreversible illness,
3. increased serious incapacitating reversible illness, or
4. a substantial present or potential hazard to human health or the environment if improperly managed (This is the basic criterion.)

More specific criteria include the following:

1. It exhibits any characteristic (I, C, R, E).
2. It is acutely toxic or acutely hazardous (H).
3. It is otherwise toxic (T).

The criteria for listing hazardous waste are for use by EPA. However, a generator's waste may still be a hazardous waste under Section 7003 (imminent hazard) or 3007 (inspections) of the Act. If EPA has reason to believe that the waste is hazardous within the definition (i.e., the basic criterion) of hazardous waste—Section 1004 (5) of the Act, it can be so designated. Additional criteria for listing acutely toxic or hazardous are as follows:

1. Small doses cause human mortality.
2. Oral LD_{50}= <50 ppm by weight (mg/kg).
3. Inhalation LC_{50} = <2 mg/l.
4. Dermal LD_{50} = <200 mg/kg.

5. Otherwise caused and increased mortality, serious irreversible illness or serious incapacitating reversible illness. According to EPA, the toxicity criterion is based also on the fact that if a waste contains a hazardous constituent listed in Appendix VIII (of RCRA) it is presumed to be hazardous.

Listing of Appendix VIII items can be argued to the contrary by providing the following information about the hazardous constituent:

1. nature of toxicity,
2. concentration.
3. migration potential,
4. persistence or degradation potential,
5. quantities,
6. plausible types of improper management,
7. past damage cases,
8. regulations by other federal or state agencies, or
9. other appropriate factors.

Substances will be listed in Appendix VIII only if they have been shown in scientific studies to have toxic, carcinogenic, mutagenic or teratogenic effects on humans or other life forms. Appendix VIII lists 361 items, the sources of which include the following:

1. Priority Toxic Pollutants—307(A) Clean Water Act — (CWA)
2. Hazardous Pollutants—112 Clean Air Act — (CAA)
3. Known Carcinogens—Cancer Assessment Group—(CAG)
4. Hazardous Materials—311 Clean Water Act (CWA)
5. Poisons A&B—Department of Transportation (DOT)
6. Pesticides (canceled)—Federal Insecticide, Fungicide and Rodenticide Act (FIFRA)
7. Acute Toxics—National Institute for Occupational Safety and Health (NIOSH)
8. Acute Hazardous—Dangerous Properties of Industrial Materials (DPIM) by N. Irving Sax

Lists of Hazardous Wastes—Nonspecific Sources

These hazardous wastes are numbered F001-F016 and include basically the following categories:

1. Spent Solvents
2. Electroplating Wastes
3. Metal Heat-Treating Wastes
4. Mineral Metal Recovery Wastes
5. Air Emission Scrubber Sludges
    (a) Coke oven wastes
    (b) Blast furnace wastes

EPA has indicated the basis for listing wastes by applying the hazardous codes (I, C, R, E, H and/or T). Appendix VII of the regulation identifies the constituent that caused EPA to list wastes in Sections 261.31 and 261.32, i.e., "F" and "K" coded waste listings, respectively.

Lists of Hazardous Wastes—Specific Sources

These hazardous wastes are numbered K001–K069 and include the following categories:

Specific Sources	Process Wastes
1. Wood Preservation	Sludges
2. Inorganic Chemicals	Sludges and residues
3. Organic Chemicals	Still bottoms, distillation residues, spent catalysts, other residues
4. Pesticides	Sludges
	Filter solids
	Still bottoms
	Distillation residues
5. Explosives	Sludge
	Spent carbon
6. Petroleum Refining	Slop oil
	API separater sludge
	Tank bottoms
7. Leather Tanning	Various solid waste
	Sludges
8. Iron and Steel	Pickle liquor
	Emission control dust/sludge
	Ammonia still lime sludge
9. Primary Copper	Acid plant blowdown sludge
10. Primary Lead	Sludge
11. Primary Zinc	Sludge
12. Secondary Lead	Emission control sludge

Each hazardous waste listed is assigned an EPA number. This number must be used in complying with the notification, recordkeeping and reporting requirements.

Lists of Hazardous Wastes—Commercial Products

This list applies to 361 commercial chemical products that are discarded or intended to be discarded. This list includes two separate lists: 239 toxic wastes (T) and 122 acute hazardous wastes (H). Both listings apply to the commercial chemical product itself, as manufactured; off-specification products; and residues and debris of spills on land and water. The commercial chemical product refers to a chemical substance that is manufactured or formulated for commercial or manufacturing use only. It does not refer to process waste that contains any of toxic or hazardous substances listed. Process wastes deemed hazardous waste by EPA will be listed in the specific and nonspecific sources either because these wastes contain a commercial chemical product or are identified as hazardous waste by testing characteristics. The acute hazardous waste listing also includes containers used to hold any of the commercial products listed and liners removed from containers used to hold these products, unless these containers and liners are triple rinsed with a suitable solvent or by some method shown by tests or acceptable literature to achieve successful removal of the chemical.

Delisting Standards

Delisting standards apply only to listed waste from individual generators or their wastes discharged from a hazardous waste management facility. The generator must demonstrate that his waste is fundamentally different from that listed; i.e., it must not exhibit the characteristics of ignitability, corrosivity, reactivity or extraction procedure toxicity using applicable test methods.

For acute hazardous waste, a generator must demonstrate that his waste does not meet the following criteria: (1) an oral LD_{50} of less than 50 mg/kg (rat); (2) an inhalation LC_{50} of less than 2 mg/l (rat); and a dermal LD_{50} of less than 200 mg/kg (rabbit). Also, a waste can be demonstrated as nonhazardous if EPA concludes that the waste is nonhazardous based on the absence of any toxic constituent listed in Appendix VIII and knowledge about the waste including: nature of toxicity, concentration, migration potential, persistence, bioaccumulation, quantity, severity of damage, plausible types of improper management and other regulatory actions. For a generator to delist a toxic waste, however, he must demonstrate that his waste does not contain any toxic constituent given in Appendix VIII or show that the above factors can knowledgeably argue against the waste being hazardous.

The delisting procedure involves submitting the information required (i.e. proposed action, justification and supporting information) to the EPA Administrator in accordance with Section 260.20 of the regulations. EPA may then publish the submission, grant the delisting proposal, deny the delisting proposal or revoke the delisting proposal. If the delisting proposal is granted or denied, it will be published in the *Federal Register* and provided a public hearing based on the EPA administrator's discretion.

CONCLUSION

Can the RCRA system work? Hazardous waste can be managed properly by people working together, not by regulations alone. Success can be realized only if the states, EPA regions, the regulated communities, EPA headquarters, and the citizens of the United States work in cooperation toward a common goal.

We need to move away from situations like the Love Canal by developing, constructing and using techniques such as high-temperature incineration. We need to move away from disasters like the fire at Elizabeth, New Jersey by using better storage methods.

We need to move away from open dumping practices for unknown wastes by using properly designed, constructed and operated landfills and by maintaining the identity of the waste, tracking it and keeping appropriate records. Managing hazardous waste properly is definitely a challenge of the 1980s that confronts all of us now.

CHAPTER 2

HAZARDOUS WASTE: HISTORICAL AND ETHICAL PERSPECTIVES

P. Aarne Vesilind
Department of Civil Engineering
Duke University
Durham, North Carolina

We speak glibly and confidently about "hazardous wastes" as if we knew exactly what they are and recognized without reservation the need to control them.

In point of fact, there has been and remains honest disagreement about what is meant by the term "hazardous," and there are serious unanswered questions about our responsibilities in the use and disposal of materials that subsequently may cause environmental damage or have an adverse effect on health.

In this chapter I hope to address first the question of what is a "hazardous waste" and present this discussion in historical perspective. Second, I would like to consider our responsibilities in hazardous waste management, not from the legal and administrative sense, but from the ethical.

WHAT IS A HAZARDOUS WASTE?

It wouldn't be difficult to be snared in a semantics trap in trying to define "hazardous," since this word is not freestanding. It requires qualifiers, objects and explanations. Hazardous to whom? Under what circumstances? At what concentrations and at what frequencies? These and many other questions must be answered, at least operationally, in order to proceed.

One can argue that every material, under given circumstances, can cause disease, suffering or damage, and should thus be classified hazardous. Water, for example, is certainly a threat to life and property if it comes as a flood.

At the other extreme, we can argue that any substance, at a low enough concentration or in a proper circumstance, will *not* cause discernible ill effects. Radiation, used properly, certainly has a net positive value in the diagnosis and cure of human diseases. Some of the heavy metals (e.g., zinc, copper) are necessary nutrients in the human metabolism. We cannot live without *some* of these in our diet. The definition of hazardous must be within these limits—all materials or none of them.

Historically, the problem with hazardous substances was related to industrial toxicology. Before the industrial revolution it was only at the workplace that the public encountered materials at sufficiently high concentrations to result in health problems.*

Probably the best and earliest description of such health effects is the classic work by Georgium Agricola, *De Re Metallica*, originally published in 1556 [1]. Agricola, whose real name was Georg Bauer, was born and raised of humble origin but attained wide respect as a physician, teacher, historian and "Bürgermeister" in his native Bohemia in the midst of the most prolific metal mining district in Central Europe. A contemporary and friend of Erasmus, he is now noted mainly for his studies of mining, including the diseases of the miners.

For example, he writes in one chapter:

> If the dust (in the mines) has corrosive qualities, it eats away the lungs, and implants consumption (tuberculosis) in the body; hence, in the mines of the Carpathian Mountains, women are found who have married seven husbands, all of whom this terrible consumption has carried off to a premature death. Further, there is a certain kind of *cadmia* (arsenical cobalt) which eats away the feet of the workmen when they have become wet, and similarly their hands, and injures their lungs and eyes.

In another section he describes the use of fire-setting for liberating ores from rock in which the fire is built underground. He cautions that:

*There are, of course, exceptions, such as the problem with the high lead content in the drinking water of the noble Romans who could afford to have lead pipes for bringing water to their homes.

> While the heated veins and rock are giving forth a foetid vapor and the shafts or tunnels are emitting fumes, the miners and other workmen do not go down in the mines lest the stench affect their health or actually kill them.

Figure 1 is an original rendering showing a workman leaving the tunnel after setting the fires.

Agricola was followed in early writings about occupational health by a number of authors, but none had the impact of Bernardino Ramazzini, an Italian physician who published a thoughtful book, *Diseases of Workers*, in 1700. Ramazzini not only surveyed occupational diseases, but suggested potential measures to combat them and called for laws to make the

Figure 1. Original rendering showing a workman leaving a tunnel after setting the fires.

workplace safer. His work was translated into many languages and remained as a definitive volume for well over 100 years [2].

During the 18th century, however, little was done to control the effect of hazardous materials in the workplace. Effects of various chemicals took on the names of the trades with which they were associated. Probably the most famous occupational disease was mercury poisoning, suffered by the hatters. Hats had been made from a felt that was produced by "sizing" soft hairs; pressing them into felt. This process was found to be greatly enhanced by the use of mercuric nitrate. The "mad hatters" who performed this task were suffering from chronic mercury poisoning [3].

Benjamin Franklin, an apprentice printer, learned about the dangers of lead [4]. In a letter to a friend in 1786 he wrote:

> When I was in Paris with Sir John Pringle in 1767, he visited La Charite, a Hospital particularly famous for the cure of that Malandy, and brought from thence a Pamphlet containing a List of the Names of Persons, specifying their Professions or Trades, who had been cured there. I had the Curiosity to examine that List, and found that all the Patients were of Trades, that, some way or other, use or work in Lead, such as Plumbers, Glaziers, Painters, etc., excepting only two kinds, Stonecutters and Soldiers. These I could not reconcile to my Notion, that Lead was the cause of that Disorder. But on my mentioning this Difficulty to a Physician of that Hospital, he inform'd me that the Stonecutters are continually using melted Lead to fix the Ends of Iron Balustrades in Stone; and that the Soldiers had been employ'd by Painters, as Labourers, in Grinding of Coulours.

He concluded, after presenting other examples:

> You will see, that the Opinion of this mischievous Effect from Lead is at least above Sixty Years old; and you will observe with Concern how long a useful Truth may be known and exist, before it is generally receiv'd and practic'd on.

Laws to curb the more blatant abuses of worker safety were long in coming. In 1877 the state of Massachusetts passed the first such statute [5]. In 1910, in the face of countless mining disasters, the U.S. Congress created the Bureau of Mines.

These actions did not prevent the continued loss of life, however, and the prevalence of occupational disease, as illustrated by the vents at Gauley Bridge, West Virginia in 1930 and 1931. The New Kanasho Power Company, a subsidiary of Union Carbide, was to build a hydroelectric power plant to supply electricity for a Union Carbide facility and had to build a tunnel through a high silica rock mountain near Gauley Bridge. The Rhinehart-Dennis Company, the contractor, recruited black workers from nearby states for the arduous job, paying them 50¢ per hour. The working and living conditions were deplorable. According to a study of the tragedy [5]:

> The dust was often so thick that workers couldn't see ten feet in front. Though the West Virginia mining law required a thirty-minute wait after blasting, workers were herded back into the tunnel immediately after the blast. A foreman at times had to beat them with pick handles to get them to return.

The silica content was extremely high, and soon workers began dropping dead. Through an arrangement with a local undertaker, the time between death and burial was three hours, thus preventing autopsies. In all, 169 black workers ended up in that graveyard, 2 or 3 to a hole.

After the magnitude of this disaster became known, the U.S. Public Health Service investigator reported to a congressional committee in 1961 that 476 people had been killed and 1500 disabled during the construction of the tunnel [5].

Such disasters prompted a government crackdown on hazardous materials in the workplace, culminating in the passage of the Occupational Safety and Health Act (OSHA).

In 1970 hazardous wastes were still not considered to be a public nuisance or an environmental health concern. For example, the 1970 edition of the annual report by the Council on Environmental Quality doesn't even *mention* hazardous wastes.

Then, one by one, local health problems due to old abandoned chemical dump sites began to surface, and suddenly everyone was concerned with hazardous wastes. But no one had yet defined what was *meant* by a hazardous waste. This vacuum created a flurry of activity by numerous government agencies and private groups, all suggesting appropriate definitions. A summary of the factors included by a number of these definitions is shown in Table I.

The system developed by Battelle can be considered as representative of these efforts and is summarized in Figure 2 [6]. Interestingly, all of these definitions were developed between the years 1972 and 1974 [7]. Before that, the dictionary definition of hazardous as "marked by danger; perilous" had to suffice.

It was hoped by some that the uncertainty in what is meant by a hazardous material would be resolved by the wording of the Resource Conservation and Recovery Act (RCRA). The RCRA definition of a hazardous waste is as follows:

> A solid waste, or combination of solid wastes, which because of its quantity, concentration, or physical, chemical or infectious characteristics may cause . . .

Table I. Classification Criteria

System	Criteria											
	Toxicological	Flammability	Explosivity	Corrosivity	Reactivity	Oxidizing Material	Radioactivity	Irritant	Strong Sensitizer	Bioconcentration	Carcinogenicity, Mutagenicity, Teratogenicity	Sufficient Quantity
Title 15, U.S. Code, Sec. 1261	X	X		X			X	X	X			X
CPSC-Title 16, CFR, Part 1500	X	X		X			X	X	X			X
Food, Drug, and Cosmetic Act	X									X	X	X
DOT-Title 49, CFR, Parts 100-199	X	X	X	X		X	X	X				X
Pesticides-Title 40, CFR, Part 162	X	X								X	X	
Ocean Dumping-Title 40, CFR, Part 227	X									X	X	X
NIOSH-Toxic Substances list	X										X	
Drinking Water Standards	X									X	X	
FWPCA Sec. 304(a)(1)	X									X	X	
FWPCA Sec. 307(a)	X									X	X	X
FWPCA Sec. 311(b)(2)(A)	X											X
Clean Air Act-Sec. 112	X							X			X	
California State List	X	X	X	X				X	X			
National Academy of Sciences	X	X			X							
TRW Systems Group	X	X	X	X			X			X		X
Battelle Memorial Institute	X	X	X		X	X	X	X		X	X	
Booz-Allen Applied Research, Inc.	X	X	X		X							X
Dept. of the Army	X											
Dept. of the Navy	X	X	X	X	X	X	X					
National Cancer Institute	X									X	X	X

> increase in mortality or an increase in serious illness . . . or . . . pose a substantial present or potential hazard to human health or the environment when improperly treated, stored, transported, or disposed of, or otherwise managed.

Unfortunately, this definition is woefully inadequate. Not only is the word "hazard" used in the definition of "hazardous" but words such as "serious," "substantial," "improper," etc. beg for definitions. In legal terms this definition is perhaps adequate since shades of meaning can be ironed out in court. For engineers and scientists, however, this is not an operational definition.

Recognizing this dilemma, the EPA established four characteristics as yardsticks by which to judge the degree of hazard:

- ignitability
- corrosivity
- reactivity
- toxicity

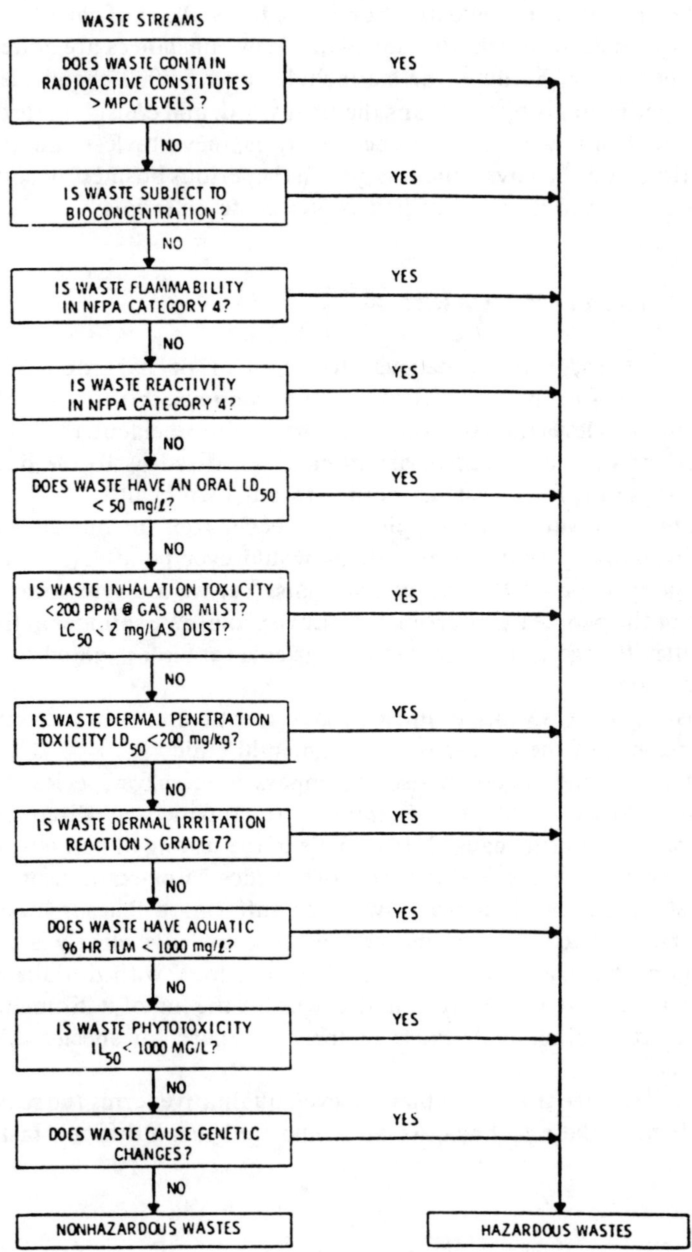

Figure 2. Graphic representation of the hazardous waste decision model.

all determined by special extraction procedures. A list of chemicals falling under this definition was developed and new substances are added to the list as necessary. To date, the list includes over 55,000 chemicals.

This definition is by no means the final word, and considerable debate is presently being waged to change it. It is, nevertheless, an operative definition, and the government is now in the serious business of controlling the handling and disposal of hazardous waste materials.

WHY SHOULD WE CARE?

Most of us, I trust, believe that control of hazardous waste by government is proper and necessary. We recognize this first on no higher philosophical level than self-survival. The Kepone incident, the Love Canal and other well-publicized disasters have sensitized us to the dangers of improper hazardous waste disposal—to our own health.

There is, however, in my opinion, a deeper root for our concern. It is actually unlikely that any one of us would ever be affected directly by improperly handled hazardous materials. Yet we are sympathetic to the plight of the people living around the Love Canal. We shudder at the events of Gauley Bridge and admire Georg Agricola for his farsighted concern for worker safety.

This altruism is a basic human characteristic and is in fact one of the cornerstones of the emerging environmental ethic.*

But environmental ethics also encompass concern for species other than humans. Most of us, for example, are appalled at discovering the widespread suffering caused to animals in the testing of new cosmetics, or the use of animals for the furthering of "science." For example, it may be of interest to the psychologist how much suffering a Rhesus monkey can stand before becoming psychotic, but the knowledge clearly is not worth the agony experienced by the monkey. We look with disdain on those "sportsmen" who shoot eagles and hawks for the fun of it. Somehow these people are subhuman because of their disregard for species other than humans.

And yet, if we tried to explain in even qualitative terms the reasons and logic behind these feelings, we all would find this a difficult task.

*Ironically, altruism also a major impediment to population control, as pointed out by Alastair Gunn [8]. If we didn't care for the welfare of our fellow man, there would be significantly fewer people populating the globe.

Part of the problem of explaining our feelings arises from the inability to apply moral philosophy of the past to our present problems. For example, the utilitarian philosophy of Bentham and Mill argues that what is right is what brings the greatest pleasure. Modified utilitarianism takes into account the greater value of intellectual or spiritual goods, as opposed to worldly pleasures. Such comfort in intellectual pleasures may appeal to some, but would not make a total human being.* Can a poem be as lovely as a tree? Or as valuable? And does the tree have its own standing [9]?

The categorical imperative philosopher Immanuel Kant suggests that certain actions such as lying, stealing and killing are *always* wrong and one ought to live one's life so as not to engage in such immoral activities; a clearly impossible commandment.

The Judeo-Christian traditions offer little help. Lynn White [10] has argued convincingly that these traditions are in fact antienvironmental and promote overpopulation and despoilment of the environment.

Eastern religions, while professing to promote harmony in nature, are in fact anthropocentric. The reincarnation doctrine places a value on animal life, but only because of the *human* souls that may be temporarily inhabiting the animals.

Thus, little seems to be gained from classical and modern moral philosophy to explain our feelings toward the environment. There are no doubt useful and thought-provoking ideas that emerge from the tradition of exotic sources, such as the American Indian or the Finno-Ugri peoples of northern Europe. But these ideas can only act as stimulants in the development of our own environmental ethic [8].

We cannot, at this time, explain why we care. Our reasons cannot be totally selfish because we often have nothing personally to gain from our feeling of concern. Neither can we argue that we are doing this for our progeny, since as the famous one-liner puts it, what has our progeny ever done for us? We are thus in a quandary and have yet to wiggle our way out.

Of course, not all of us share the same values. Consider the man who rented the farm in Toms River and used it for toxic waste disposal. Over 140 wells were contaminated with petrochemicals. Was what he did unethical? He may have acted ethically by his standards. Were he here today, he might defend his actions as "laissez-faire capitalist entrepreneurship." He certainly didn't break any laws, so should he be considered a criminal?

---

*H. G. Wells, in his *War of the Worlds*, speculates about the possibility of humans becoming cattle for the invading martians who eat humans for food. Well fed and cared for in the barnyard, could humans adapt to being domesticated?

Another approach to environmental ethics is to use a "body count." The 140 wells is not really a serious matter and the dumping was not unethical since no one was killed. To people who espouse this ideal, the Kepone incident is an overblown media spectacular, since no one was directly killed by the chemical [11]. The damage to wildlife and the legacy of a tainted estuary and permanently contaminated land, willed to future generations to clean up, is not to them unethical.

In short, we do not agree on the ethics of hazardous wastes. What is "right" and what is "wrong" becomes a very personal decision. Translating the "is" to "ought" is even more difficult. Much remains to be done in this area of ethics.

CONCLUSION

It is, in retrospect, much easier to define what we mean by hazardous waste than it is to define why we should be concerned about hazardous waste. In fact, it is fairly simple to get everyone to agree that Kepone, a chlorinated hydrocarbon pesticide, decachlorooctahydro-1,3,4-metheno-2H-cyclobuta (cd) pentalen-2-one, should be classified as a hazardous waste than it is to explain why we should be concerned in the least about the fathead minnow, the blue crab or the Canadian geese that inhabit the James River estuary.

REFERENCES

1. Agricola, G. *De Re Metallica*, translated from the first Latin edition of 1556 by H. C. Hoover and L. H. Hoover (New York: Dover Publications, 1950).
2. Teleky, L. *History of Factory and Mine Hygiene* (New York: Columbia University Press, 1948).
3. Hamilton, A., and H. Hardy. *Industrial Toxicology* (New York: Paul B. Holber Inc., 1949).
4. Smellie, W. G. *Public Health: Its Promise for the Future* (New York: Macmillan Publishing Co., Inc., 1955).
5. Page, J. A., and M-W. O'Brien. *Bitter Wages* (New York: Grossman, 1973).
6. Program for the Management of Hazardous Wastes for Environmental Protection Agency," Final Report, Battelle Memorial Institute, Richland, WA (1973).
7. Kohan, A. M. "A Summary of Hazardous Substance Classification Systems," U.S. EPA SW-171, Office of Solid Waste Management Programs, Washington, DC (1975).
8. Gunn, A. *Environmental Ethics* Duke Environmental Center Publication, Duke University, Durham, NC (1980).

9. Stone, C. D. *Should Trees Have Standing?* (Los Angeles: William Kaufman Inc., 1974).
10. White, L. "The Historical Roots of Our Ecologic Crisis," *Science* 155: (March 10, 1967).
11. Bender, M. E., and R. J. Huggett. "Kepone in the St. James River," *Environ. Sci. Technol.* 14:8 (1980).

CHAPTER 3

PUBLIC HEALTH IMPACT OF HAZARDOUS WASTE DISPOSAL

Avram Gold

Department of Environmental Sciences & Engineering
School of Public Health
University of North Carolina
Chapel Hill, North Carolina

An assessment of the public health impacts of hazardous waste disposal must include a definition of the scope of the problem. A breakdown of the current rate of waste generation [1] is given in Table I. Of this enormous amount of waste, 80% is disposed of at the site of generation. Only an estimated 10% is disposed of according to existing regulations. The amount of material involved, the wide dispersal of disposal sites and the lack of adequate disposal sites indicate a potential for serious adverse impacts.

Problems arise because the chemical compounds in the waste do not remain fixed where they are deposited. Some have appreciable vapor pressure or are gaseous at ambient temperatures (such as vinyl chloride trapped in PVC process sludges) and will diffuse through fill, appearing in ambient air at the disposal site. Ground water percolating through site boundaries moves chemicals along the direction of flow. Even relatively insoluble compounds can be slowly moved because of equilibrium between adsorbing particles and the moving ground water (similar to processes involved in chromatography), resulting in mass migration of the contents

Table I. Estimate of Current Rate of Waste Generation

Waste Generated	Amount (millions of tons)
Organic Chemicals	11.7
Primary Metals	9.0
Electroplating	4.1
Inorganic Chemicals	4.0
Textiles	1.9
Petroleum Refining	1.8
Rubber and Plastics	1.0
Miscellaneous	1.0

of the disposal site. Wells in the path of migrating chemicals (leachate) will become contaminated and the ground water eventually may feed surface waters, contaminating ponds, lakes and streams. Some of the impact on public health is obvious: (1) community air pollution from volatile components carried beyond perimeters of disposal sites, and (2) pollution of ground water supplying wells and surface waters used for drinking and recreation. Current events are replete with examples of human exposures brought about through these mechanisms.

THE TRACK RECORD

A Lathrop, California, Hooker Chemical facility processed waste water in an unlined lagoon [2]. The water contained residues of fertilizer, nitrates and dibromochloropropane, a suspect carcinogen known to have produced sterility in workers involved in its manufacture and packaging. Water from the lagoon filtered through the permeable basin and into the water table, eventually poisoning wells on adjoining property that were used to water cattle.

A Wilmington, North Carolina landfill, operated with state approval in the early 1970s by Waste Industries, contaminated ground water supplying wells in the nearby community of Flemington [3]. Odor and foul taste have made water unpotable and, for the foreseeable future, residents will have to obtain drinking water from tank trucks.

A landfill in the vicinity of Happauge, Long Island [4] was used to dispose of sludge from a vinyl chloride polymerization process by the approved method of layering sludge, garbage and earth [5]. However, vinyl chloride, a known carcinogen producing angiosarcomas of the liver, diffused to the surface and was detected in concentrations as high as 90 ppm in air vented from the dump. The presence of vinyl chloride forced the closing of a neighboring elementary school.

During routine sampling of beef fat in its meat and poultry inspection program, the U.S. Department of Agriculture (USDA) discovered some samples that contained up to 1.5 ppm hexachlorobenzene. The samples were traced to cattle grazing in pasture land around Darrow and Geismar, Louisiana, where it was discovered that 100 mi^2 of pasture were contaminated by hexachlorobenzene that had vaporized from a pit disposal site [4].

The most celebrated incident, and a situation that will certainly receive continuing attention in followup studies is, of course, Love Canal [6]. The impact of this disposal site on public health will be meticulously documented. At Love Canal, leachate migrated from the disposal site into adjacent properties and infiltrated basements of dwellings. Saturated foundation walls became inexhaustable reservoirs of noxious vapors in which the known leukemogen benzene and a number of aliphatic and aromatic chlorinated hydrocarbons have been identified. Leachate that collected in basements was removed by sump pump and drained into a storm sewer system that discharged into the Niagara River and Lake Ontario. Both the river and lake are sources of public drinking water and are used for recreational fishing, so contamination of these waters exposed a large population to potential effects from the waste chemicals.

Benzene and chlorinated hydrocarbons were identified in ambient air at a school built on the former disposal site. Before the acute nature of the problem became apparent, topsoil had been carried away from the site by area residents and spread over gardens, further distributing the waste through surrounding residential areas.

The hazardous wastes that caused the problems described above were legally disposed of at sites approved by the appropriate state and local agencies. Far greater potential for damage is created by illegal disposal.

A stretch of 270 miles of North Carolina highway [4] was saturated with PCB from used transformer fluid by a trucker who opened the valves on his tanker and discharged his cargo of waste along the highway shoulder in the small hours of the morning.

A contractor rented a farm at Toms River, New Jersey [7] under the pretext of establishing a drum reclaiming facility; in fact, he used the site to illegally dispose of 4500 55-gallon drums containing wash solvents and waste organics from Union Carbide. The owners of the farm discovered the dump as a result of odors emanating from the property. Two years after the removal of the drums, a rash of complaints by surrounding homeowners regarding taste and odor of their well water resulted in the discovery of ground water contamination by the chemicals at ppm levels. As a consequence, 148 wells had to be sealed permanently, and the residents will have to live with the anxiety of potential illness from an exposure that cannot be defined with respect either to duration or concentration.

Chemical Control Corporation of Elizabeth, New Jersey filled a waterfront warehouse with 40,000 drums of crystalline picric acid and nitroglycerine, both explosives, and bacterial and low-level radioactive wastes. On discovery of the hazard, New Jersey, fearing the results of ignition of the waste, declared a state of emergency pending cleanup. However, the site did catch fire and a disaster of major proportions would have occurred but for the offshore breeze.

In a vacant lot in the Bronx, New York, a trucker illegally dumped refuse from a doctor's office; the refuse contained strychnine tablets, cyanide salts and mercury [8]. Fortunately, the material was discovered and removed before neighborhood children who played in the area were injured.

EFFECTS OF HUMAN CONTACT

Effects of human contact with hazardous waste components can be classified into two categories. The first is short-term effects, usually involving high levels of exposure. The second is chronic effects, resulting from long-term, lower-level exposure. With the exception of Love Canal, none of the incidents described above have involved documented injury—yet. Unfortunately, there are examples of both serious injury and death resulting from massive short-term exposures to hazardous wastes.

In most cases, the short-term effects can be related directly to toxic action of specific chemicals involved. For example, oil contaminated with tetrachlorodibenzodioxin (TCDD), one of the most toxic organic chemicals known to man, was inadvertently spread for dust control on three racetracks and a farm road in Missouri [4]. As a result, 63 horses and numerous pets and wild birds died. A child who played on the farm road required hospitalization with severe bladder pains and urinary bleeding.

Many other examples exist. In Perham, Minnesota, a contractor unknowingly sank a well on a site where arsenic-based insecticides had been buried 30 years previously [5]. Eleven workers who drank the contaminated water suffered severe arsenic poisoning. At the Love Canal site, children and adults who came in contact with leachate developed dermatological problems, presumably chloracne, from the chlorinated hydrocarbons present as major components of the waste. The driver of a truck in Iberville, Louisiana, was asphyxiated while discharging waste into a disposal pit when the contents of his load reacted with waste already in the pit.

Long-term effects are much more difficult to identify. Low-level exposures to complex mixtures of chemicals are involved, so toxicological experience, usually available only for pure compounds, may not be at all relevant in identifying exposure-related injury. Further compounding the problem is the difficulty in reliably estimating the duration of the exposure and the concentration of the chemicals involved. Because of relatively small populations at risk in particular incidents, statistically significant increases in widespread types of health problems, such as respiratory illness, miscarriages, birth defects and common forms of cancer, are very difficult to demonstrate. Only for very large rate increases or aggregations of rare illnesses can a clear-cut relationship between exposure and effect be made. Two examples of rare tumors that led to the rapid identification of carcinogenic exposures were mesotheliomas from exposure to asbestos and angiosarcomas of the liver from exposure to vinyl chloride. An increase in the rate of stomach or breast cancer, however, would have been extremely difficult to detect. A plethora of more subtle, nonquantifiable effects often ascribed to chronic exposures—behavioral and personality changes, increased suicide rates, neurological problems and anxiety—are even more difficult to confirm. Often they are reported only after discovery of a hazardous waste problem and the attendant publicity. This is the case at Love Canal, where "psychophysiological" complaints such as depression, irritability, dizziness, nausea, weakness, fatigue, insomnia and numbness in the extremities have been prevalent since discovery of the dump. Anecdotal evidence, such as the assertion by a local physician of an elevated rate of respiratory illness in a neighborhood adjacent to the site, has been noted but is extremely hard to verify. Interestingly, another chemical disposal site in Elkton, Maryland [4] is located within 400 feet of a residential area where there have been recent complaints of sore throats, dizziness and respiratory problems similar to those accompanying the development of the Love Canal situation. At the Elkton site, however, contact of residents with leachate has yet to be demonstrated.

RECENT STUDIES

Some patterns are beginning to emerge from the intensive studies surrounding Love Canal, with the caveat that the results are still tentative and that the correlations indicated are statistical, with no direct causal relationships demonstrated. Under the guidance of Beverly Paigen of Roswell Memorial Park Institute, geographical occurrences of miscarriages, birth defects and respiratory illness have been correlated with swales (dry stream beds intersecting the canal) that may provide routes for migration of leachate [6]. The results of this study apparently have been supported in an expanded study conducted by the state as a result of Paigen's initial findings; however, the New York State Department of

Health has been unwilling to disclose its raw data. The most recent panel on the controversial chromosome damage study conducted for the U.S. Environmental Protection Agency (EPA) by Biogenics, Inc. of Houston viewed the raw data and concluded that there were, in fact, types of chromosome damage present that are rare in "normal" individuals and, for this reason, advised that the study should be followed up by a more rigorously designed survey. The implications of excessive chromosome damage (should excessive damage eventually be demonstrated) are that the exposed population will be at a higher risk of developing cancers than a nonexposed population of similar ethnic and socioeconomic background [10].

Another chronic exposure to leachates from a chemical waste disposal site occurred near the town of Legeler in southern New Jersey near the U.S. Naval Air Station at Lakehurst [11]. Since some time after commencement of disposal operations in 1972, the population has been exposed through contaminated well water to benzene, trichloroethylene, dichloroethylene and tetrachloroethane. Excessive kidney problems have appeared among the population, ranging in severity from sudden and repeated occurrence of kidney stones in individuals with no personal or family history of this problem to major loss of renal function, requiring dialysis. Although the causal relationship has not been rigorously proven, renal problems are an effect that would be expected from chronic exposure to chlorinated hydrocarbons. Kidney and liver ailments have been potential problems identified among the population at Love Canal, although statistically significant rate increases cannot be confirmed.

An aspect of public health impacts of hazardous waste disposal that had not received much consideration until Love Canal is the psychological problems arising from the stress of coping with events and anxiety that future serious illness may stem from past exposure. The stress is amplified by lack of any routine government procedures for dealing with such situations. Some results of stress are already evident at Love Canal [12]. Effects on marriages have been dramatic: of 237 families involved in the 1978 evacuation, 40% have undergone separations or divorces. Intracommunity relationships have undergone strikingly opposite changes. The need for psychological self-help in a generally aloof blue-collar community suspicious of mental health professionals has resulted in dissolution of many longstanding feuds and an increase in socializing.

In summary, acute effects on health have been documented and, in many cases, the nature of the physiological problems is consistent with the exposure to the known contaminants. Long-term health problems have not been studied adequately, but effects undoubtedly will involve a range of physiological systems—genetic, hepatic and respiratory. Psychological effects have not received attention until recently, but for large-scale incidents such as Love Canal, they appear serious and costly.

REFERENCES

1. *EPA J.* 5:12 (1979).
2. *The New York Times* (August 6, 1979).
3. *The News and Observer*, Raleigh, NC (July 6, 1980).
4. *Environmental 80/81, Annual Editions* (Guilford, CT: M.H. Brown, 1980), p. 175.
5. Crossland, J. *Environmental 80/81, Annual Editions* (Guilford, CT, 1980), p. 100.
6. Emper, L. R. *Chem. Eng. News* 58:22 (1980).
7. Maugh, T. H. *Science* 204:819 (1979).
8. *The New York Times* (August 9, 1979).
9. Davis, P. *EPA J.* 5:14 (1979).
10. Kolata, G. B. *Science* 208:1239 (1980).
11. *The New York Times* (February 7, 1980).
12. Holden, C. *Science* 208:1242 (1980).

CHAPTER 4

DECISION-MAKING IN HAZARDOUS WASTE DISPOSAL

George W. Pearsall
Department of Mechanical Engineering
Duke University
Durham, North Carolina

There are almost as many ways to make decisions as there are decision-makers. In this chapter, I will not even attempt to describe how decisions actually are made in hazardous waste disposal; already there have been too many examples of decisions based on intuition, political expediency, snap judgments, simple beliefs, rule following and other, even more irrational, procedures. Rather, the analysis presented here will be limited to rational procedures of decision-making: to the characteristics that make them rational, to a methodology for analyzing these procedures and determining the values contained within them, and to an assessment of whether the final decision is congruent with that set of determined values.

DECISION-MAKING PROCEDURES

Most rational decision-making procedures can be subdivided into five stages.

Identification of the Problem

The first stage is identifying the problem. Superficially, for individuals dealing with hazardous waste problems this usually is done by others, e.g., a telephone call or a memo about PCB dumped along North Carolina highways, discovery of an illegal dump in Rhode Island, ground water contamination around Love Canal in New York, vinyl chloride monomer in the air around a sludge dump on Long Island, and so on.

Because of the potential dangers (often reinforced by political or economic pressure) of hazardous wastes, decisions often must be made quickly. In spite of this temporal constraint, at the first stage it still is important to state, concisely yet completely, what the principal issue is in the problem [1]. Are people in imminent danger? If so, how many? How serious is the danger? Is the hazard primarily to the environment? Is the principal issue one of public relations rather than danger? Is the hazard acute (short term) or chronic (long term)? Is the problem an allegation, the confirming details of which must still be obtained in order to verify it? Can the danger be measured? Are some important characteristics of the danger not quantifiable?

I will assume, for the analysis presented in this chapter, that the problem is a recently discovered site into which hazardous wastes have been dumped in the past, that scientific data confirm a definite hazard that represents danger to some segment of the population or the environment, and that the decision is "What to do next?"

Determination of What Values are Important

The second stage in such a decision-making process involves determining what values are important. This is essentially an ethical issue; it must be resolved beforehand because there is seldom time for the luxury of philosophical debate when a decision has to be made, on the spot, about what to do with a hazardous waste dump site. For hazardous wastes, a reasonable starting point for this second stage might be the list of criteria proposed by the U.S. Environmental Protection Agency (EPA) in its promulgation of regulations under the authority of the Resource Conservation and Recovery Act (RCRA) [2]. A substance is considered by EPA to be hazardous if exposure to it results in:

1. increased mortality;
2. increased serious irreversible illness;
3. increased serious incapaciting reversible illness; or
4. substantial present or potential hazard to human health or the environment when it is improperly treated, stored, transported, disposed of or otherwise managed.

This is essentially a list of negative values: potential occurrences we fear or dislike. Ethically, such negative values are considered to be "bad." For example, No. 4 in the above list might be expanded to include other specific negative values common in American society today:

1. pain,
2. disfigurement,
3. tumors,
4. genetic change,
5. family dislocation,
6. property destruction,
7. animal injury or death, or
8. plant injury or death.

Naturally, these negative values have positive complements: qualities we cherish and hold dear (life, liberty, happiness, health, truth, beauty, knowledge, property, naturalness, and so on). Ethically, such positive values are considered to be "good." It is important at this point for the decision-maker to consider which values are most important and to assess any incongruency between his/her sets of positive and negative values and those of society; conflicts between values create the problems of ethics and the dilemmas of decision-makers. This latter point will be reconsidered during the discussion of the fifth stage of decision-making.

Assessment of the Hazards

The third stage entails assessing the hazards—determining in what ways and how seriously the hazards threaten our positive values or promote our negative values. The current EPA interpretation of RCRA indicates that a substance can be considered hazardous if its hazardous characteristic can be measured by an available standardized test method. Four specific characteristics have been identified:

1. ignitability;
2. corrosivity;
3. reactivity; and
4. toxicity.

Further, each of these hazards is defined in quantitative terms [2] (also see Chapter 1). The quantification of hazards is not a straightforward exercise. Typically, it requires somewhat subjective judgments about the severity of the hazard, the probability of exposure to the hazard, and the population at risk. For hazardous materials, these judgments entail answering such questions as the following: How easily ignited it is? How corrosive (and to what) is it? How reactive (and with what) is it? How toxic is it? In what concentrations? What are its modes of entry into the environment, and what are its transport mechanisms after entry? Once these questions are answered a more formal hazard assessment is possible.

Definitions

The terms hazard, danger, risk and safety are not defined uniformly in all the fields in which they are used, so I will state my working definition of each term.

Hazard: a condition that can be expected to cause damages, including injury or death to exposed individuals. Examples are: explosion, high voltage, high concentrations of a toxic chemical, fire, high levels of ionizing radiation, etc.

Danger: the potential consequences of exposure to a hazard, including some measure of severity. Examples are: death, burns, para- and quadraplegia, heart stoppage, brain damage, narcosis, birth defects and tumors. The potential for injury to some living organism usually is implied by the word *danger*.

Risk: the probability of exposure, coupled with the severity of the consequences. *Risk* often is used in a more general way than *danger* is used in that *risk* can refer to financial loss or property damage, as well as to personal injury or environmental damage. In its simplest formulation (when death is the only consequence considered), risk may be taken to be equal to the probability of exposure: "The risk of dying in an automobile accident in the United States is about 25/100,000 people per year" ($p = 2.5 \times 10^{-4}$/yr). When the hazard can result in damage or injury less severe than death, risk often is equated with the product of probability (p) and severity (S) for a given level of damage:

$$R = (p)(S)$$

A number of severity indices are available by which S can be quantified [3,4], but none has achieved universal acceptance for risk analysis calculations. A more general formulation of risk involves plotting the probability, $p(>S)$, of damage with severity, S, or greater vs the severity, S (a so-called "risk curve"). Such an integral probability curve can be reduced to a single number (equivalent to $R = (p)(S)$) by calculating the *expected value of damage*, E(D). The "expected value" of a series of mutually exclusive consequences is the sum of the product of each value (its severity,

S, in this case) with its probability, p. Replacing the sum by an integral for a continuous distribution of consequences:

$$E(D) = \int S \frac{d}{dS} p(>S)dS$$

Note, however, that reducing the risk curve to a single number entails a loss of information, as "many vastly different curves with enormously different significance could have the same expected value" [5].

Safety: the situation that exists when the risk is known and understood and is acceptable to the population at risk [6]. Note that safety does not necessarily imply zero risk.

Acceptable Risk

The actual *hazard assessment* must be done by the decision-makers, and part of this assessment must include determining what degree of risk is acceptable. The decision-makers must determine "How safe is safe enough?" The answer to this question regarding hazardous waste continues to evolve, as indicated by EPA efforts to develop tests for the following categories of hazards:

1. Radioactivity
2. Infectiousness
3. Phytotoxicity
4. Mutagenicity

Further, the identification of a waste as "hazardous" (even though its hazardous characteristic) cannot necessarily be measured by an available standardized test method if it is generated from certain sources, such as hospital departments and particular industrial processes [7].

Reviewing the set of values identified in the second stage of decision-making provides a useful check on the hazards determined in the third stage. Applying this check to hazardous wastes, one might predict that tests will be developed eventually for teratogenic and carcinogenic substances, as well as for wastes that might produce irreversible changes in the environment if present in sufficiently high concentrations (CO_2 and the "greenhouse effect") or convertible to armaments (nuclear proliferation [8,9].

Identification and Invention of Options

The fourth stage in this decision-making scenario comprises identifying and inventing options for disposing of the discovered hazardous waste. This stage places a premium on technical knowledge and creativity. Can the waste be disposed of onsite or must it be removed to an offsite location? Can it be converted, e.g., by incineration, to a less hazardous form? Can it be reprocessed for use as a feedstock? Must it be dug up or can the site be capped and sealed? Is deep well injection or landfilling (with or without cementation) possible? Will monitoring stations be required? Are other disposal options conceivable that have not been attempted yet (such as shooting it into outer space)?

An important requirement at this stage is to estimate the probability of failure, and the severity of the consequences of failure, for each option considered. In other words, a second set of hazard assessments must be performed during this fourth stage—similar to those performed during the third stage but this time for each option—to estimate the risk (including the population put at risk) of exercising each option. Failure probabilities may be estimated by standard procedures, such as fault-tree analysis (FTA) and failure-mode and effects analysis (FMEA) [10,11]; severities chosen will depend on the value system used. Both failure probability and severity can only be "engineering estimates," but it is important that they be addressed explicitly to provide the decision-maker with as much useful information as possible. If private contractors are retained for developing some disposal options, it would seem reasonable to expect from them rational estimates of failure probabilities and consequences for each option proposed.

Decision-Making

The fifth stage is making the decision. In a rational decision-making process, the decision is made by comparing the options and choosing the best option available. But what is "best?" By what criteria is "best" defined? This is essentially an ethical question and can be addressed most conveniently in terms of some of the major ethical traditions [12–14].

Hedonism

One of the oldest ethical traditions is that of hedonism; it was hedonism that was espoused by the sophists, with whom Socrates debated the nature of the good life. Hedonists adopt the concept that "pleasure is the supreme good," together with its corollary that injustice is better than justice because it brings more pleasure. Hedonistic waste disposal, for example, might be described as "Dispose of the waste anyway you can to maximize profits; if you do something illegal, don't get caught." Although there is much evidence that we live in a hedonistic society, public outrage over abandoned hazardous waste sites also indicates that we place a limit on hedonism when it affects the lives, health and liberty of others.

Plato offered an alternative to hedonism—a life in which goodness is governed by reason and in which virtue is knowledge. According to Plato, the souls of all individuals comprise three elements: reason, spirit (passion) and appetite (desire). Harmony ("good" decisions) is achieved when the soul is governed by reason, the desire is regulated by reason, and reason is supported by passion. This state of harmony is achieved only by study and diligence, and the argument led Plato to the concept of the "just man," who was transformed—in a decision-making context—to the "philosopher-king." In other words, final Platonic decisions on hazardous waste disposal should be made, not by the masses, not in the marketplace, not by pressure from the media, but by a just person who has achieved great knowledge by study and diligence. We can trace some of our ideas of professionalism to these roots and they beg the questions: Are we educating engineers broadly enough to become the "philosopher-kings" of hazardous waste management? Are our political decision-makers educated deeply enough to make truly just decisions? Have the complexities of our modern, technological, bureaucratic world precluded the usefulness of Plato's ideas?

Categorical Imperative

A third alternative is that of the categorical imperative, advocated by Kant. A categorical imperative is a principle with which every person should act consistently, with *no* exceptions, with the emphasis here on duty (thou shalt not lie [15], thou shalt not kill) rather than on the ends to be achieved by the action. A Kantian hazardous waste disposal policy might state, for example, that every possible effort will be taken to avoid exposure of any living organism to hazardous materials and that the decision-makers will be honest with the public. Note that cost is not taken into account in such a policy; neither is the possibility of other kinds of risks created by following the moral imperative. The Occupational Safety and Health Administration's (OSHA) pursuit of the proposed benzene standard [16] might be considered a Kantian approach to a toxic hazard.

Utilitarianism

An alternative to the categorical imperative is to focus on the ends achieved by any action or decision. The utilitarianism of Bentham, later modified by Mill, embraces this shift in focus and usually is translated as "the greatest good for the greatest number." In the utilitarianism of Bentham, "good" was equated with pleasure and "bad" was equated with pain or the privation of pleasure. Mill broadened the definitions so that a "good" utilitarian decision was one that maximized happiness and minimized unhappiness; he also allowed for differences in the quality of pleasures. All people are equal in utilitarianism, though; your pleasure and pain count as much as mine. Most current decision strategies (decision analysis, cost/benefit analysis, risk/benefit analysis [17] and other offshoots of operations research) trace their ethical origins to utilitarianism. Before blindly following such a "utilitarian calculus," however, it is desirable to evaluate the values on which it is based. Are they congruent with your personal values? Are they congruent with society's values? Where are the value conflicts and how can they be resolved?

The above four traditions—hedonistic, Platonic, Kantian and utilitarian—appear to me to be the most ingrained in present attitudes toward hazardous waste disposal. But one also can identify some other approaches to the problem (not all of which possess an extensive "tradition" behind them) that merit at least a brief mention.

Legalism

A fairly common approach to making decisions can be described under the heading *legalism*, by which I mean "do whatever the law requires, but nothing more." Legalism equates obeying the law with making good decisions. But what is the "law?" As a minimum, it is a set of statutes that prohibit certain actions, plus a set of judgments that have been handed down on cases—from which one can infer (usually with the assistance of an attorney) what one should do to avoid both criminal and civil charges. But the "law" can also be interpreted more broadly to include codes or standards that an industry or profession has adopted to guide the behavior of its members. For example, a legalistic interpretation of my ethical responsibility as an engineer might entail only obeying the law and the Rules of Professional Conduct for North Carolina professional engineers (since the state of North Carolina has licensed me to practice).

Populism

Another approach to making ethical decisions is represented by the idea of *populism*; let the people decide for themselves. As our communication systems become more pervasive, we may actually develop the capacity for an entire population to vote by telecommunication on an issue such as what to do with a hazardous waste site, after having been presented with the alternatives. For the moment, though—with the exception of issues so important as to warrant a referendum—the populism advocates must rely on some representative group, which claims to be expressing the wishes of the people. Some contenders for this representative role in hazardous waste management appear to be the U.S. Congress, EPA, various consumer activist groups and the communications media. A "good" decision, according to this modified populist scenario, would be whatever decision best satisfied the group that was representing the people's interest.

Situation Ethics

Another ethical decision-making strategy, which has a long tradition, might be called the *golden rule* strategy: "Do unto others as you would have them do unto you." In other words, put yourself in the victim's shoes and make decisions as you would want them made, were you the victim. An extension of this approach has been popularized in the last twenty years under the heading of *situation ethics*, in which the essence of "good" behavior is to act congruently with your love for others [18].

Retributive Justice

At another pole from the golden rule is an even older tradition—one that I call *retributive justice*: "An eye for an eye, and a tooth for a tooth." In other words, do to others what they would do (or have done) to you when you are (were) in their circumstances. It is worth pointing out that retributive justice is considered a marked improvement over what it replaced, which was typically "A life for an eye, and an eye for a tooth."

Altruism

Among the more recently popular ethical philosophies, one also finds *altruism*, the beliefs and associated behavior that are based on the idea of putting consideration for others ahead of your own selfish interests. Altruism is the opposite of *egoism*, or selfishness, which in turn may be viewed as a somewhat more benign version of hedonism.

Existentialism

Also among the more recent philosophies with ethical implications in hazardous waste disposal is *existentialism*: the belief that man's existence is central, that an understanding of the world derives from that centrality. (As the T-shirt on one of my students expressed it, "I think; therefore I am—I think.") According to the existentialist, there is no grand scheme, no overall direction to the world. If one views life (or waste disposal) as an existentialist "card game," not only does the number of cards change each hand, but over the course of a few hands the suits change, and so do the rules. The parallel with the operations of some federal agencies seems obvious.

Skepticism

The last ethical tradition I choose to mention here may be one of the oldest, yet often it is neglected in discussions of ethics. It is *skepticism*, the doctrine that all knowledge is uncertain. In its extreme form, it is the belief that nothing will work (or that anything attempted will make things worse). Murphy's law: "If anything can go wrong, it will," epitomizes skepticism. Skepticism usually leads to (or rationalizes) inaction. In its modern American form, it sometimes seems to derive more from knowledge than from ignorance. For example the hazardous waste skeptic may be very knowledgeable about all that can go wrong when one attempts to clean up a hazardous waste dump site. As a result of such knowledge one may be led to "educated incapacity": knowing so much about what can go wrong that one is incapable of taking any action at all.

Decision Procedures

Of what use is even a superficial knowledge of such ethical traditions to the decision-maker dealing with a hazardous waste site? It is not to provide a methodology one can follow blindly. I think the principal purpose of such knowledge is to sensitize the decision-maker to the values underlying his/her decisions. A clear comprehension of these values seems necessary for resolving the conflicts inherent in difficult decisions.

For example, if you are inclined toward cost/benefit, or risk/benefit analysis, realize that this decision technique derives philosophically from utilitarianism. If you consider yourself a utilitarian, can you defend killing an innocent to improve the "happiness" of many others by some small amount? If not, by what criteria did you decide that such killing was not defensible? (A scenario often used in ethics courses to illustrate this point involves the decision to lynch an innocent person to keep a rioting crowd from going berserk and killing even more people to avenge a murder.) A utilitarian hazardous waste analog might involve the decision to expose a very small segment of the population to a concentrated hazardous chemical (in a dump site) to avoid exposing a very large segment of the population to that same chemical in dispersed form.

On the other hand, if your decision procedures are based on the notion that *no* risk is acceptable, e.g., that hazardous materials must have concentrations below the present limits of detection, realize that this notion is a variation of Kant's categorical imperative. (One of the scenarios sometimes used to criticize this general doctrine focuses on the imperative, "Never lie under any circumstances." The decision-maker is asked the whereabouts of someone he knows to be innocent, so that person's pursuer can locate and kill him.) A hazardous waste analog might involve the question of obeying a regulation that generally produces "good" results, but which in this one instance would harm some people, e.g., by eliminating their water supply, with no concomitant benefits. Under what conditions is a decision-maker warranted in disobeying such a regulation?

In addition to the conflicts between rule-following decision philosophies (e.g., Kantian) and end-promoting philosophies (e.g., utilitarian), other conflicts invariably arise. How important are the opinions of an uninformed, but frightened, public? Should those put at risk by a decision have a controlling input into the decision? If one bases decisions on love for others, how does one make a decision when love for one conflicts with love for another? A 19th century dilemma sometimes posed to illustrate this latter conflict is that of a mother's anguish over having to choose between smothering a colicky, crying baby in a wagon train attempting to sneak through a western territory at night or having the noise attract the Indians.

Beyond these "local" conflicts, a whole range of potential "global" conflicts arise when considering policy decisions. Typically, these arise from the constraint of fixed resources. For example, should government spend money to clean up a hazardous waste site when that same money would probably save more lives if it were invested in mobile coronary care units [19]? Different ethical philosophies may lead to different answers.

If everyone does not espouse the same values and does not use the same decision strategies, where can the decision-maker look for guidance? In the absence of a common cultural philosophy, my advice is as follows: First, ascertain that the choices presented to you are in fact the best choices. The creative decision-maker can invent new alternatives when faced with conflicting constraints, much as the design engineer invents new products when performance objectives appear to conflict. Remember the Hegelian dialectic in philosophy, in which the conflicts between thesis and antithesis are surmounted by rising to a new level and creating a *synthesis* [20]; the engineering parallel is striking. Second, act in such a way that you can defend your decisions *both* to a jury of your political peers (as on a jury in a court, should you be sued for malpractice or charged with breaking the law) *and* to a jury of your professional peers (as on a licensing review board or membership committee hearing, should your profession or professional society charge you with unethical conduct). Third, while you are formulating a decision that is defensible to both hypothetical juries, do so assuming "perfect discovery" of all information. That is, assume that both juries will have complete access to *all* the information to which you had access (plus that to which you *could* have had access) and that they can be educated to understand the consequences of your actions. Fourth, remember Murphy's law.

REFERENCES

1. Behn, R. D., and J. W. Vaupel. *Analytical Thinking for Busy Decision Makers* (Basic Books).
2. *Federal Register* 45(98):33121 (May 19, 1980).
3. Fine, W. T. "Mathematical Evaluation for Controlling Hazards," *J. Safety Res.* 157 (December 1971).
4. Consumer Product Safety Commission. *Neiss Data Highlights*, Hazard Identification and Analysis (published quarterly).
5. Okrent, D. "A General Evaluation Approach to Risk-Benefit for Large Technological Systems and its Application to Nuclear Power," UCLA Report, UCLA-ENG-7777 (December 1977)
6. Lowrance, W. W. *Of Acceptable Risk* (Los Angeles: William Kaufman, 1976).
7. Hanrahan, P. "Hazardous Wastes: Current Problems and Near-Term Solutions," *Technol. Rev.* 20 (November 1979).
8. Ramsay, W. *Unpaid Costs of Electrical Energy* (Baltimore: Johns Hopkins Press, 1979).
9. Schurr, S. *Energy in America's Future: The Choices Before Us.* (Baltimore: Johns Hopkins Press, 1979).
10. Society of Automotive Engineers. "ARP 926A, Fault/Failure Analysis Procedure" (1979).
11. Atallah, S. "Assessing and Managing Industrial Risk," *Chem. Eng.* 94 (September 8, 1980).
12. Albert, E. M., T. C. Denise and S. P. Peterfreund. *Great Traditions in Ethics* (New York: D. van Nostrand Co., 1969).
13. MacIntyre, A. *A Short History of Ethics* (New York: Macmillan Publishing Co., Inc., 1966).
14. Melden, A. I. *Ethical Theories: A Book of Readings* (Englewood Cliffs, NJ: Prentice-Hall, Inc., 1967).

15. Bok, S. *Lying: Moral Choice in Public and Private Life* (New York: Pantheon Books, Inc., 1978).
16. "Benzene Verdict: A Knell for Pending OSHA Rules?" *Chem. Eng.* 73 (August 11, 1980).
17. "Risk/Benefit Analysis in the Legislative Process," Joint Hearings before the Subcommittee on Science and Technology, U. S. House of Representatives, No. 71, July 24, 25, 1979, U. S. Government Printing Office, Washington, DC (1980).
18. Fletcher, J. *Situation Ethics* (Westminster Press, 1966).
19. Acton, J. P. "Evaluating Public Programs to Save Lives," Rand Report, R-950-RC (1973).
20. Pirsig, R. *Zen and the Art of Motorcycle Maintenance* (New York: Bantam Books, Inc., 1974).

PART 2

TECHNOLOGICAL INPUTS TO THE MANAGEMENT PROCESS

CHAPTER 5

HAZARDOUS WASTE INCINERATION

Harry Freeman

Research Program Manager
Incineration Research Branch
IERL
Cincinnati, Ohio

Several things now happening in the country are increasing the attractiveness of high-temperature incineration as a means of disposing of hazardous industrial wastes. Increasingly stringent landfill regulations and rising public objection, even to suitable landfills, have sharply reduced industry's traditional method of disposing of wastes; concern for the effects of industrial wastes on biological sewage treatment processes and tighter environmental regulations have eliminated the sewer system as an acceptable outlet for many wastes [1]. Finally, the passage in 1976 by the Congress of the Resource Conservation and Recovery Act (RCRA) and the Toxic Substances Control Act (TOSCA) began the process by which additional environmental regulations will continue to increase the attractiveness of incineration [2]. This chapter will explore several of the questions surrounding hazardous waste incineration and discuss possible roles for incineration as a part of the nation's approach to improving the management of its industrial wastes. Included are discussions of (1) current incinerator technology; (2) two specific incineration systems, and (3) pending questions surrounding the RCRA incineration regulations.

THE WASTES

Each day American industry produces enough waste to fill the New Orleans Superdome floor to ceiling five times [3]. The U.S. Environmental Protection Agency (EPA) estimates that about 90% of these wastes can be classified as hazardous using the RCRA criteria for determining whether a waste is hazardous; this amounted to over 57 million metric tons in 1980 [3]. Of this amount, it is estimated that no more than 6% is presently being disposed of through controlled incineration [4].

The greatest number of toxic and hazardous wastes are organic or partially organic [2] and consequently are suitable for destruction through high-temperature incineration, either with or without the use of supplementary fuel. For a waste to sustain combustion, it has to have a calorific value of at least 8,000–10,000 Btu/lb and, if a liquid, be pumpable. Materials that fall into this category are light solvents (toluene, benzene, acetone and ethyl alcohol) and heavy organic tar and still bottoms similar to residual fuel oil [5]. Liquid organic combustible wastes are usually atomized into a high-temperature furnace fired on an auxiliary fuel and the combustion products passed through a secondary combustion chamber of some sort at 1200° C or more to ensure complete combustion. Solids are charged directly into incinerators, which have long residence times, to ensure complete combustion [2]. Of course, wastes vary in their suitability for incineration. There exist many lists of waste which can be consulted to determine the combustibility of wastes and the mutual compatibility of wastes when they are being burned [5].

Incinerator Technology

Of the several types of hazardous waste incineration systems in use, two are generally cited as being the most versatile and useful—liquid injection systems and rotary kilns. Some incineration facilities are combinations of these two types of units.

Liquid Injectors

Liquid injection incinerators are versatile units that can be used to dispose of virtually any combustible liquid that can be pumped. There are a wide variety of such units presently in operation throughout the country [6]. The units can be either horizontal or vertical. The major components of a system include one or more liquid and/or auxiliary fuel burners, a primary air blower, one or more secondary air blowers, and an emergency relief stack in addition to the refractory line combustion chamber itself. Liquid injection incinerators operate at temperatures between 820° C and 1600° C (1500–3000° F). Gas-phase residence times range from 0.1 to 2 seconds [5]. A flow diagram for a liquid injection system is shown in Figure 1, and a list of wastes currently burned in liquid waste incinerators is shown in Table I. Advantages and disadvantages of liquid injection systems are shown in Table II.

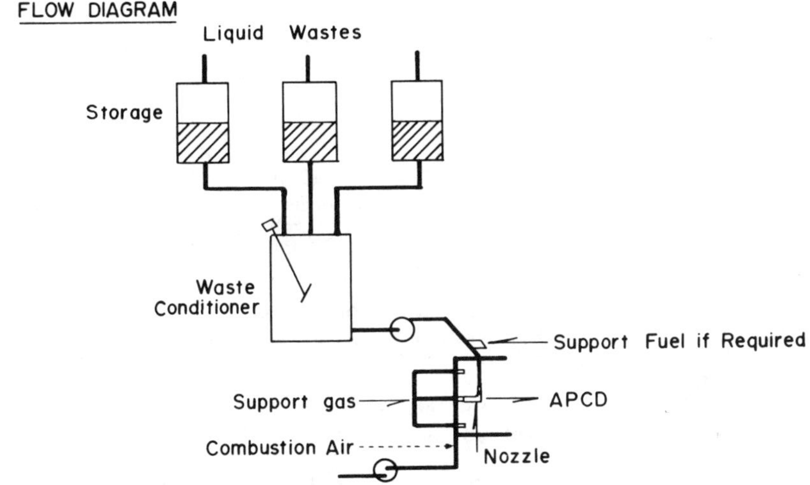

Figure 1. Liquid injection flow diagram.

Rotary Kilns

Rotary kilns are versatile units that can be used to dispose of solids, liquids, slurries and gaseous combustible wastes [6]. A typical rotary kiln/afterburner system is shown in Figure 2. Rotary kilns are long, cylindrical rotating furnaces lined with firebrick or other refractory in which solids are combusted by themselves or incinerated by combustion of an auxiliary fuel or liquid wastes [5].

Rotary kilns are used to handle large volumes of both solid and liquid wastes. They can be designed to handle contained wastes. Rotary kilns are especially effective when the size or nature of the wastes preclude the use of other types of incineration equipment. Combustion temperatures range from 870° to 1600° C (1600° F to 3000° F), depending on the waste material

Table I. Hazardous Wastes that are Burned in Liquid Injection Incinerators

Phenols	Hexachlorocyclopentadiene
Still and reactor bottoms	Organophosphate pesticides
Cyanide and chrome plating wastes	Waste from polymer polyol production
Polyester paint	Dodecyl mercaptan wastes
Polyvinyl chloride paint	Fluorinated herbicide wastes
Thinners	Ethylene glycol manufacture residue
Solvents	Waste residues from alkyl benzene production
Off-specification isoprene	Perchloroethylene manufacture still bottoms
	Alkyl and oryl sulfonic acid wastes
	Still bottoms from acetaldehyde production
	Nitrochlorobenzene

Table II. Liquid Injection Incineration—Advantages and Disadvantages

Advantages

1. Capable of incinerating a wide range of liquid wastes.
2. No continuous ash removal system is required other than for air pollution control.
3. Capable of burning smaller amounts of waste than the nominal design rate.
4. Fast temperature response to changes in the waste fuel flowrate.
5. Virtually no moving parts.
6. Low maintenance costs.

Disadvantages

1. Must be able to atomize liquids through a burner nozzle or other device except for certain limited applications.
2. Heat content of liquids must maintain adequate temperatures or a supplemental fuel must be provided.
3. Must provide for complete combustion and prevent flame impingement on refractory.
4. Burners susceptible to pluggage, depending on wash characteristics.
5. Sophisticated instrumentation required to maintain efficient combustion.

combustion characteristics. Solids residence times vary from seconds to hours, depending on the type of waste [5]. Advantages and limitations of the rotary kilns are shown in Table III.

Wastes currently incinerated in rotary kilns are shown in Table IV.

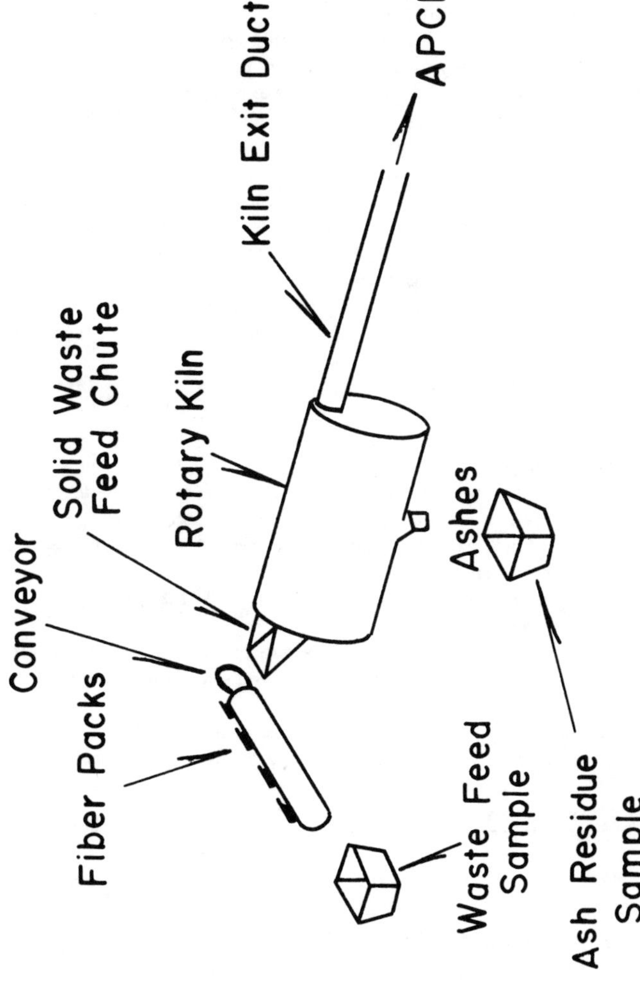

Figure 2. Rotary kiln flow diagram.

Table III. Rotary Kilns—Advantages and Disadvantages

Advantages

1. Will incinerate a wide variety of liquid and solid wastes.
2. Capable of receiving liquids and solids independently or in combination.
3. Not hampered by materials passing through a melt phase.
4. Feed capability for drums and bulk containers.
5. Wide flexibility in feed mechanism design.
6. Provides high turbulence and air exposure of solid wastes.
7. Continuous ash removal that does not interfere with the waste oxidation.
8. No moving parts within the kiln.
9. Adaptable for use with a wet gas scrubbing system.
10. The retention of residence time of the waste can be controlled by adjusting the rotational speed.
11. The waste can be fed directly into the incinerator without any preparation such as preheating, mixing, etc.
12. Rotary kilns can be operated at temperatures in excess of 1400°C (2400°F), making them well suited for the destruction of toxic compounds that are difficult.
13. The rotational speed control of the kiln also allows a turndown ratio of about 50%.

Disadvantages

1. High capital cost for installation, especially for low feedrates.
2. Operating care necessary to prevent refractory damage.
3. Airborne particles may be carried out of kiln before complete combustion.
4. Spherical or cylindrical items may roll through kiln before complete combustion.
5. Kiln incinerators frequently require excess air intake to operate due to air leakage into the kiln via the kiln end seals and feed chute, which lowers supplementary fuel efficiency.
6. Drying or ignition grates, if used prior to the rotary kiln, can cause problems with melt plugging of grates and grate mechanisms.
7. High particulate loadings.
8. Relatively low thermal efficiency.

THREE APPROACHES TO INDUSTRIAL WASTE INCINERATION

The Liquid/Fluid Incineration Facility—Metropolitan Sewer District (MSD) for Greater Cincinnati, Ohio

Since 1979 the Cincinnati MSD has operated a facility capable of incinerating up to 40,000 gpd of liquid industrial wastes and grease and grit skimmings from a municipal sewage facility. The facility, which is the only

Table IV. Rotary Kilns Applicable Wastes

Epichlorohydrin manufacturing wastes
Steam still bottoms from aniline and alkylated phenol production
Acrylonitrile manufacturing wastes
Reactor tar bottoms from adiponitrile manufacture
Phenolic tar from 2,4-D manufacture
Chlorotoluene production wastes
Phenylamine tar wastes
PCB wastes in capacitors
Evaporate residue from the cumene process for phenol manufacture
Nitrochlorobenzene tars
Catch basin grease, nitrile pitch from production of surface active agents
TDI manufacture reactor tar bottoms
Mercaptobenzothiazole tars
Polyvinyl chloride

such high-temperature incinerator operated by a public institution in the country (Figure 3), includes a rotary kiln, cyclone furnace, combustion chamber, high-energy venturi scrubber, induced draft fan and a stack. Flue gases from the kiln and cyclone furnace flow into a common combustion chamber. The flue gases from the combustion chamber enter the venturi scrubber and then discharge through the fan and the stack to the atmosphere [7].

The facility presently accepts liquid waste only from tank trucks and maintains several large holding and blending tanks to blend the wastes as necessary to ensure safe combustion. Of paramount concern in any waste management operation is the cost. A schedule of prices charged by the Cincinnati MSD facility is shown in Table V.

The EPA Mobile Incineration System

The EPA, through its Oil and Hazardous Materials Spills Branch in Edison, New Jersey is developing a mobile incinerator which can be transported to spill and disposal sites to provide onsite disposal of hazardous chemical wastes [8,9]. The system is intended to ease the environmental and political problem presented by having to transport wastes to central incinerators.

The system, designed to incinerate highly refractory compounds containing chlorine and phosphorus such as tetrachlorodibenzodioxin (TCDD), polychlorobiphenyls (PCB) and Kepone, which may be in pure form, in sludges, or in soils, is self-contained on three aligned semitrailers, with only water and fuel to be supplied externally. The system includes a rotary kiln, a secondary combustion chamber and gas stream processing

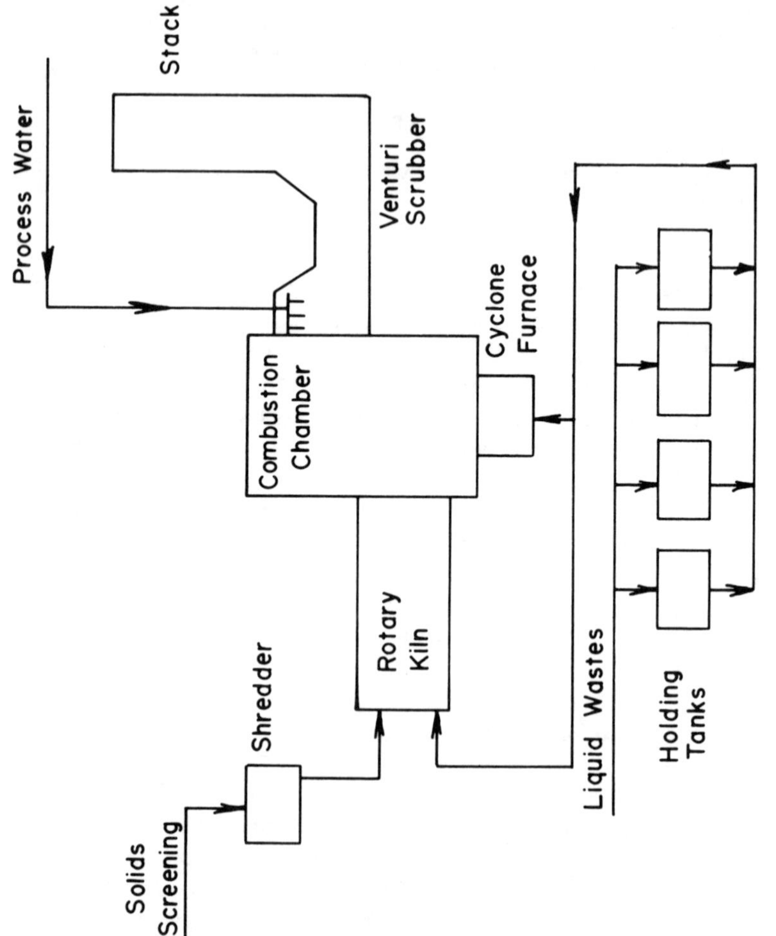

Figure 3. Cincinnati MSD.

Table V. Classification of Liquid Fluid Wastes at Cincinnati MSN L/F Incinerator

General Description	Materials Included	1977 Survey (% of Total)
Light hydrocarbons and nonaqueous solvents	Benzol, toluol, aromatics, cellosolve	10
Medium- to heavyweight hydrocarbons, etc.	Crankcase oils, still bottoms, transformer oils	5
Low water content aqueous wastes	Clabberstock, soaps, fatty acids, alcohols, cutting oils	10
Dirty solvents	Kerosene, soluble oil residue, oil-soluble inks, ink wastes, organic pigments	36
High water content aqueous wastes, semisolids and sludges and low heating value liquids	Paint overspray, liquid polymers, chlorinated solvents, oil sludge	19
Skimmings from Mill Creek WTP	Grease, soaps, etc.	20

equipment. The system meets standards for transport on interstate highway systems. The processing capacity of the system is 75 gal/hr of #2 diesel fuel or material of equivalent Btu content, 7000 lb/hr of dry slightly contaminated sand or 170 gal/hr of contaminated wastes.

The unit is presently undergoing shakedown tests at Edison and will be available to support the needs of EPA offices concerned with hazardous and toxic wastes.

Rollins Environmental Services, Deer Park, Texas

The Rollins Deer Park Facility is located about 20 miles east of Houston in a highly industrialized area near the Houston ship channel. It has been operating since 1971 and operates 24 hours a day, 7 days a week. When operating, the unit incinerates approximately 250,000 pounds of waste per day consisting of bulk liquids, solids and slurries.

The incineration system includes a rotary kiln and a liquid injection burner feeding into a common afterburner. Solid wastes are conveyer fed into a rotary kiln and incinerated at temperatures of from 1200–1500°F. Combustion gases are passed through an afterburner to attain temperatures of 2300–2400°F. Estimated total residence time is 2.6 seconds. Ash from the kiln is disposed of onsite in an approved landfill.

A venturi scrubber is used to remove particulate from the exhaust gases. Lime is injected to neutralize the scrubber water in a single pass system. The used scrubber water then enters settling ponds and is further treated. The exhaust gases exit through a 100-foot stack.

During 1976 the EPA sponsored some test burns at the Rollins Facility to evaluate the effectiveness of thermal destruction of two different industrial wastes: (1) discarded electrical capacitors containing PCB and (2) waste from the production of nitrochlorobenzene. Among the findings of the tests were that high-temperature incineration in rotary kilns is an effective disposal method for the subject wastes, with destruction efficiencies exceeding 99.99% for every test except the incineration of whole capacitors which was 99.5%.

ENVIRONMENTAL REGULATIONS

Proposed Regulations, December 18, 1978

The Resource Conservation and Recovery Act authorizes the EPA to establish regulations for the incineration of hazardous wastes. The proposed regulations for incineration of hazardous wastes were published in the *Federal Register* on December 18, 1978. They placed on owners/operators a number of performance requirements coupled with various operating standards, most of which were intended to help ensure that the performance criteria were met regularly.

The performance criteria required that incinerators burning hazardous wastes achieve a destruction efficiency of 99.9% or better, and a combustion efficiency of at least 99.9%; that particulate emissions be less than 270 mg/scm^3 (0.12 gr/scf) at zero excess air; that fugutive emissions be controlled; and emission controls remove more than 99% of the halogens when hazardous wastes containing more than 0.5% halogens were burned.

Proposed operating regulations required that owners/operators monitor and record significant variables at 15-minute intervals; that trial burns be conducted, analyzed and reported to the regional administrator before each new and "significantly different" hazardous waste was incinerated; and that the wastes be retained for 2 seconds at 1000°C combustion temperature (1200°C for halogen-containing wastes) and more than 2% excess oxygen (3% for halogen-containing wastes). A "note" or variance provided that incinerators need not comply with the detailed combustion criteria if an equivalent combustion efficiency could be achieved by other means. A device to automatically cut off waste feed whenever combustion or scrubber conditions changed significantly was a final requirement.

The EPA received comments on the proposed regulations. Many of the comments dealt with broad and general issues, such as the propriety of applying any design and operating requirements, rather than relying solely on performance criteria based on destruction. Some comments questioned the Agency's statutory authority to regulate hazardous waste incineration under RCRA at all. Conflicts with the regulation of particulate emissions under the Clean Water Act were cited.

Many comments were essentially technical. One comment suggested that turbulence criteria be added to the time and temperature requirements. Others suggested that measurement methodology was unclear or difficult to apply, and that the measurement locations and frequencies for emission and effluent temperature measurements needed to be specified. Three comments pointed to dangers inherent in automatic cutoff devices and suggested that gradual shutdowns were preferable.

A middle range of comments accepted the general framework of the proposed regulations but felt specific criteria were unnecessary or infeasible. In particular, it was claimed that the halogen destruction levels appropriate for chlorine were more stringent than those necessary or possible for other halogens, such as bromides or iodides. Similar comments suggested that the general 99.99% destruction efficiency requirement was impractical and very costly, and would divert wastes from relatively "safe" incineration into dangerous, long-term landfills. Some comments suggested varying destruction efficiency in accord with each waste's degree of hazard.

The high cost of trial burns and of trial burn analysis was mentioned frequently. There was considerable confusion about the practical impact of the requirement that a new trial burn be held before each "significantly different" new waste was incinerated. Several comments pointed out that the almost infinite variety of chemical mixtures making up wastes made this requirement both incredibly expensive and impossible to apply with certainty. A few comments focused on the vagueness of the requirement that fugitive emissions be controlled.

Finally, many comments suggested specific exemptions or additional inclusions to the regulations when finally promulgated. The most important of these dealt with clarifying the status of waste oils and solvents, of incinerator ash residue and scrubber effluents, and of cement kiln incinerators and utility boilers burning hazardous wastes for their thermal value.

May 19, 1980

Based on the comment received on the December 18, 1978 regulations, on May 19, 1980 the Agency issued the first phase of a two-phase approach for standards for hazardous waste incinerators. These regulations, recognizing that there was a need for guidance for incinerators during the interim period while data on which to base firm operating regulations were being developed, established some general operation requirements for incinerators that could be implemented during the interim status period. The second phase of the regulations, containing technical criteria for issuing permits and reflecting more of the Agency's responses to comments, were to be promulgated in 1980. There is no question that the rule-making under RCRA will continue for years as more information becomes available.

REFERENCES

1. Trapp, J. H. "Incineration for Industrial Liquid Wastes."
2. Ross, R. D. "The Burning Issue: Incineration of Hazardous Waste," *Poll. Eng.* 25–28 (August 1979).
3. Rogers, J. A. "RCRA Regs: Enforcing EPA's Most Pervasive Statute," *Legal Times of Washington* (May 12, 1980).
4. Stevens, J. et al. "Thermal Destruction of Chemical Wastes," paper presented at the 71st Annual Meeting of the AIChE, November 16, 1978.
5. Sittig, M. "Incineration of Industrial Hazardous Wastes and Sludges," Noyes Data Corporation, Park Ridge, NJ (1979).
6. Manson and Unger. "Hazardous Materials Incinerator Design Criteria," EPA-600/2-79-198 (October 1979).
7. Hescheles, C. A. and R. F. Bonner, Jr. "Ultimate Disposal of Wastes by Pyrolysis and Incineration" paper presented at the National Incinerator Conference, ASME, Miami, FL, May, 1974.
8. "EPA Develops Mobile Incineration System for Clean-Up of Hazardous Chemical Spills and Industrial Dumpster," EPA Technigram, IERL, Cincinnati, OH.
9. Tenzer, R. and W. Mattox. "Design and Testing of Mobile Incineration Systems for Spilled or Waste Hazardous Toxic Materials," Paper presented at the 1980 Conference and Exhibition on Control of Hazardous Material Spills, Louisville, KY, May 15, 1980.

CHAPTER 6

IN SITU TREATMENT/CONTAINMENT AND CHEMICAL FIXATION

Alexandra P. Wright and Steven D. Caretsky

Fred C. Hart Associates, Inc.
New York, New York

The nationwide emphasis on the cleanup of abandoned hazardous waste dump sites and hazardous waste spills has created the need for the development of methods that can be used to render these areas environmentally "safe" and/or to prevent further pollution by containing the material within the site. In situ physical containment and treatment are two methods available for use at hazardous waste dump or spill sites. In addition, a section pertaining to chemical fixation processes available for use on hazardous wastes has been included, as these processes are a rapidly developing technology for the treatment of hazardous wastes.

IN SITU PHYSICAL CONTAINMENT

In situ physical containment of hazardous wastes is a viable method of isolating surface and subsurface areas that have been contaminated from the surrounding environment.

A system to physically contain the wastes in a manner that is environmentally acceptable must include the following parameters:

- Surface runoff control
- Impermeable barriers
- Capping or surface sealing
- Monitoring
- Security
- Inspection and recordkeeping
- Safety equipment and planning
- Utilities

Surface Runoff Control

Surface runoff control measures should be implemented to divert water from entering the burial/spill site, prevent site drainage from entering the actual burial/spill area, and to receive drainage from the actual burial/spill area. The following control techniques should be considered for implementation:

- Drainage diversion channels and culverts
- Surface stabilization
- Subsurface drains
- Dikes

Drainage Diversion Channels

Surface runoff diversion involves constructing earth berms and excavating channels to direct runoff to natural drainageways downslope of the landfill. Channels may be constructed upslope to divert the runoff away from the site and to divert drainage away from the actual burial site.

Channels must be designed to handle, at a minimum, surface runoff flow from a 10-year, 24-hour precipitation event. The expected volume and velocity of runoff will determine the type of channel to be constructed. Channels can be made of earth, lined with asphalt, rubble or sod, or constructed of concrete or corrugated metal pipe.

Surface Stabilization

Stabilization of the surface soil will not prevent runoff but will control runoff velocities, thereby reducing erosion of the material covering the burial site. Well-compacted, fine-grained soils should be used for final cover to enhance runoff while minimizing infiltration. The completed burial site surface should be maintained at a slope of at least 2% to ensure that precipitation incident on the surface will run off to drainage facilities, but less than 5% to minimize surface erosion.

Subsurface Drains

A series of subsurface perforated pipes can be placed in a layer immediately above the actual burial site but below the surface. The drain system will intercept precipitation that has penetrated the cover material. Pipes should be connected to a common header that conveys the flow to a collection pond.

Diversion Dikes

A diversion dike is a ridge of compacted soil placed above, below or around the burial site to intercept runoff and direct it to a disposal area. All diversion dikes should be machine compacted and have positive drainage to an outlet. Diverted runoff that has not come in contact with the actual burial site or leachate from seeps should be routed to a settling basin or other settling structure for removal of sediment prior to discharge to a receiving stream. Diverted runoff that has been in contact with the actual burial site or waste leachate from seeps should be routed to a separate storage structure for testing and/or treatment prior to discharge to a receiving stream.

The following factors must be considered when designing a dike:

- Foundation conditions
- Stability analysis
- Soil selection
- Slope inclination
- Seepage control
- Allowance for freeze/thaw and dry/soak

Impermeable Barriers

Impermeable barriers can be constructed at the site to prevent the lateral migration of the waste from the immediate burial site and prohibit the movement of ground water into the burial site. Impermeable barriers may be constructed of any material that will block the lateral migration of water and waste solution. Barriers should extend to an impermeable layer at a depth below the waste trench to prevent vertical migration of wastes or ground water.

Grout Curtain

A grout curtain may be constructed around the perimeter of the actual burial site to contain the wastes and prohibit ground water intrusion.

The type of group suitable will be determined from the geological investigations. Accurate test borings, core sampling and laboratory analysis for porosity, hydraulic conductivity and connected pore volume must be conducted. Grouting agents used to seal porous strata include cement, epoxy resins, liquid glass, polyisocyanate, acrylic amide, chrome-lignin and urea formaldehyde.

Imperwall®

Imperwall is the trade name for a technique that involves driving an I-beam to a desired depth and grouting the void space created during the extraction of the beam. The beam has a steel pipe attached to its full length. As the beam is extracted, a cementitious sealing material is continuously injected through the pipe. A continuous barrier is created by overlapping successive injection points [1].

Surface Sealing

Surface sealing involves the construction of an umbrella cap or seal on the burial site to prevent water infiltration and minimize leachate generation. Clays, fly ash, soils, soil-cement lime-stabilized soil, membrane liners, bituminous concrete and asphalt/tar materials may be used for the construction of caps and seals.

Prior to construction of any seal, the burial site surface must be graded and compacted to provide a firm subgrade. Side slopes no steeper than 18% and top slopes of 12% maximum and 6% minimum are recommended. Soil cover should be placed on any sealing material that may be damaged by exposure to the sun.

Monitoring

Air Monitoring

An air monitoring program should be developed to meet the individual requirements of the site. Monitoring requirements for an undisturbed site will be far less demanding than one required for a site with ongoing activity (excavation).

Modeling requirements include dispersion data and are a prerequisite for station placement. Field meteorological stations of varying sophistication may be required but are dependent on available information.

High-volume samplers are strategically placed to gather representative samples of fugitive particulate emission. In addition, vapor-sensing devices may be required to analyze solvent evaporation and transport mechanisms [2].

Surface Water Monitoring

Monitoring programs such as this are dependent on definition of the following: upstream, entry point and downstream. Selection of the above require field investigation. With respect to water, one is concerned with the water column since the specific gravity and solubility have bearing on the material location within the column. In addition, bottom sediment also must be considered as a study parameter.

Suspect contaminants should be determined in advance, if possible. The alternative is to perform extensive physical, chemical and biological analyses and develop upstream and downstream profiles [2].

Groundwater Monitoring

A comprehensive ground water monitoring program must be implemented at the site to provide sampling and analysis of potential local aquifer pollution from site leachate or spills. A ground water monitoring program can detect and evaluate construction deficiencies in liners, collector systems, treatment areas and any onsite retention ponds.

Ground water monitoring programs can be either passive or active in nature. A passive monitoring system includes periodic testing at permanent stations strategically located in the path of the ground water flow to provide comparative data for evaluation against baseline data.

An active monitoring system is employed for point source contamination and the contaminated ground water is intercepted by utilizing a continuous pumping system.

Security

A security system must be provided to prevent the unauthorized entry of persons and/or livestock onto the site. The system must provide adequate surveillance of the entire site.

Recordkeeping and Inspection

A recordkeeping and inspection schedule are essential for the safe operation of an onsite storage facility. Records should contain, at a minimum the following information:

1. Waste characterization
2. Monitoring data
3. Inspection results
4. All maintenance performed

Inspection of the containment area should occur on a scheduled basis that has been deemed adequate to ensure proper containment of the waste.

Safety Equipment and Planning

1. Portable fire extinguishers, fire control equipment, spill control equipment and decontamination equipment should be available at the site.
2. A device or contingency plan is needed to summon external emergency assistance from local police and fire departments.
3. A Spill Prevention Containment and Countermeasure plan (SPCC) should be prepared following those requirements outlined in the *Federal Register* (40 CFR 112,151) [2].

Utilities

Utilities are necessary at the site to provide safe and efficient working conditions. The following utilities should be available:

- Fire water supply—water buffalo or pumper
- Electricity
- Telephone/radio communications

IN SITU TREATMENT

In situ biological, chemical and physical treatment techniques are available for use in treating hazardous waste dump/spill sites. These techniques include: (1) soil flushing, (2) chemical detoxification, and (3) microbial innoculations. These treatment techniques are used most effectively when the depth of contamination is shallow and the area contaminated is relatively small.

Soil Flushing

The soil flushing process consists of a series of shallow well points that are used to collect seepage subsequent to flooding the contaminated area. This process is most effective when the contaminant(s) is readily soluble in water. The flushing of certain contaminants can be accelerated by the use of an alternate flushing solution (i.e., slightly acidic solutions used for altering soil pH to accelerate flushing of metal hydroxides).

Elutriate collected by pumping the shallow well points must be tested to determine whether treatment is needed prior to discharge. Should treatment be required, packaged physical/chemical treatment units may be suitable for the contaminated elutriate.

Chemical Detoxification

The chemical detoxification process consists of flooding or injecting the contaminated area with a substance that will detoxify the contaminant. This process must be limited to contaminants that are easily degradable, have nontoxic breakdown products, and/or are convertible to insoluble precipitates.

Microbial Inoculations

The microbial inoculation process consists of seeding the soil in the contaminated area with a microbial population capable of metabolizing the contaminant. This process is only applicable if the contaminant is an organic compound and may be more time consuming than other restoration processes. If an immediate public health or environmental danger exists, physical and/or chemical techniques will provide more timely results [1].

ENGINEERING ALTERNATIVES FOR CONTAMINATED SITES

There are many options available for use in the restoration of hazardous waste dump sites. A typical scheme that may be used to determine the best option is to study the following alternatives (in situ treatment/containment, removal, storage and treatment) while considering public health, environmental, financial, time and technological constraints.

In Situ Containment

This approach will involve the isolation of the disposal site and the adjoining contaminated water and soil from the environment. The isolation will consider the surface and subsurface areas of the designated isolated area. Sealants range from naturally occurring clays and minerals to synthetic materials.

The application of sealants may involve borings followed by injection or excavation with backfilling. Long-term monitoring of air, surface water and ground water would be attendant requirements.

Waste Removal

This approach should be addressed if in situ containment is unacceptable. All other options are based on the excavation and removal of the waste and associated contaminated material. Included in this approach are the requirements for immediate and temporary storage onsite.

Storage

In general, short-term storage is defined as less than one year; long-term storage is greater than one year. Requirements for structural integrity and security must be considered.

The location of the storage area (offsite vs onsite) will be determined by safety, health, environmental and economic considerations. The appropriate selection must meet regulatory requirements. Federal, state and private alternatives should be considered.

Treatment

The primary concern of treatment is to destroy the contaminant waste. Three basic alternative treatment approaches should be considered: physical, chemical and biological.

Transportation

When contemplating the above options, one must consider transportation of the waste when offsite removal is suggested. Routes, methods and associated hazards and risks must be considered for this facet of the operation.

Risk Assessment

Safety, health and the environment are of the greatest importance for option selection. While the options under consideration may be economically feasible, they may present a serious risk to the population. Therefore, a thorough evaluation of each option should include a risk assessment. Parameters of concern such as population at risk, credible anticipated events, consequences, alternatives and statistical data are relevant to such an assessment.

Regulations

Implicit in the selection of options is the fundamental requirement of conformity with pertinent legislation. Regulatory review should be initiated as appropriate options are selected and deemed feasible.

Geohydrological

Geohydrological investigations should be conducted to determine the existence and/or probability of waste migration to the water table and underlying ground water flow system. To determine the existence of said flow paths, one must consider the following:

1. composition of the waste;
2. physical/chemical characteristics of the waste with respect to water and soil, i.e., miscibility;
3. composition of the soil/rocks surrounding the waste;
4. physical/chemical characteristics of the soil with respect to the waste, i.e., adsorptive/ion exchange capacity, permeability, thickness;
5. secondary porosity under the site, i.e., fractures, solution cavities, soil rock discontinuities, porous residual cherty layers;
6. depth and orientation of the underlying water tables;
7. permeability of the underlying bedrock aquifer and subsequent ground water flowrate; and
8. downgradient conditions.

CHEMICAL FIXATION

There are a number of chemical fixation techniques available for use in the ultimate disposal of hazardous wastes. The operation of all chemical fixation processes involves the mixing of "waste fixing" chemicals with the waste sludges. The resulting mixture is then allowed to solidify. Some of the fixation processes result in the formation of a matrix in which wastes are entrapped; or they bind such pollutants as heavy metals in insoluble complexes.

Chemical fixation processes are waste-specific, limited to the types of wastes that can be economically treated. Further, they vary in pretreatment requirements and cost. Techniques to be considered for application to hazardous wastes are listed below and then discussed in terms of their selective advantages and disadvantages and their process application:

- cement-based
- lime-based
- thermoplastic
- organic polymer
- encapsulation
- self-cementing
- glassification

Cement-Based Fixation Processes

The cement-based processes involve mixing hazardous wastes slurried in water directly with cement. The suspended solids become incorporated into the rigid matrix of the hardened concrete. Special cements have been developed for specific applications such as use in the presence of high and moderate sulfate concentrations and where rapid setting is desired. Cement additives are available that can be used to improve the physical characteristics and decrease the leaching losses from the resulting fixed sludge. Surface coatings are being developed to increase the strength and durability of the concrete-waste mixture. Coatings that have been investigated include asphalt, asphalt emulsion and vinyl. A polymer impregnation process has been developed to decrease the permeability of concrete-sludge mixtures.

Advantages of Cement-Based Processes

1. They are extremely effective for treating wastes containing high levels of toxic metals; the pH of the cement mixture enables most multivalent cations to be converted to insoluble hydroxides or carbonates.
2. Raw materials are abundant and inexpensive.
3. The process is well established and simple, with no specialized labor required.
4. The volume of cement added can be adapted to a wide range of moisture content of the waste.
5. Cement is very nonreactive; pretreatment is necessary only for materials that retard the setting reactions of cement.
6. Coating is available for leachate reduction.
7. A waste-concrete mixture can be used for subgrade and subfoundation material.

Disadvantages of Cement-Based Processes

1. The fixing process requires a large amount of cement.
2. There is a large increase in the weight and volume of the final product.
3. Uncoated products are susceptible to leaching, especially in mildly acidic solutions.
4. Expensive pretreatment or additives may be required for wastes containing impurities that affect settling and curing properties.
5. Alkalinity of cement drives off ammonium ion as ammonia gas.
6. Cement production is energy intensive.

Process Application

Cement-based waste fixation processes have been used extensively for the disposal of low-level radioactive waste. The process also may be applied to heavy metal sludges, heavy metal ash cakes, contaminated soils, mercury, chromium and cadmium sludges.

Most wastes that can be slurried in water and that combine with cement can be "fixed" by this process. However, wastes should be screened for materials that will delay the settling and curing time or weaken the bond structure of common portland cement. These materials include organic materials, silt, clay, coal, lignate and insoluble materials of $<74\text{-}\mu$ particle size.

This process is applicable to salts of manganese, tin, zinc, copper and lead and to sodium salts of arsenate, borate, phosphate, iodate and sulfide.

Lime-Based Fixation Processes

The lime-based processes involve the reaction of lime with a fine-grained siliceous material and water to produce a concrete. Fly ash, ground blast furnace slag and cement-kiln dust are the most common pozzolanic-type materials used for this process. These materials are waste products and have little or no commercial value.

Advantages of Lime-Based Processes

1. Abundant, low cost materials are used.
2. The process operation is simple and well established.
3. No specialized equipment or labor are required.
4. Extensive dewatering is not necessary.

Disadvantages of Lime-Based Processes

1. There is a large increase in the weight and volume of the final report.
2. Uncoated lime-fixed materials may require disposal in a landfill designed to contain pollutants from possible leaching.

Process Application

Lime-based waste fixation processes have been used extensively for calcium-based SO_X scrubber waste as typically produced from coal-fired utility scrubbers. The process also has been proven successful at the test level, for the stabilization of plating, steel mill and chemical process wastes. From these results it is believed that the process can stabilize wastes that have the potential to leach chemicals into the environment.

Thermoplastic Fixation Processes

The thermoplastic fixation process involves drying, heating and mixing the waste with a heated plastic matrix. The mixture is then cooled to solidify the mass. The ratio of matrix to waste is generally quite high (1:1 to 1:2 fixative to waste on a dry weight basis). The matrix and the dry waste must be mixed at temperatures ranging from 130° C to 230° C, depending on the melting characteristics of the material and type of equipment used [3]. Thermoplastic organic materials such as bitumen, paraffin and polyethylene are used for this process. A similar process uses an emulsified bitumen product as the thermoplastic material. This product is miscible with wet sludge and has the advantages of low-temperature mixing and low matrix:waste ratios.

Advantages of Thermoplastic Fixation Processes

1. Leachate loss rates are much lower than those of cement-based systems.
2. Dry disposal reduces the overall volume of the waste.
3. Matrix materials are highly resistant to attack by solutions or microbes.
4. Matrices generally adhere well to incorporated materials.
5. Materials can be extracted from the matrix if needed.

Disadvantages of Thermoplastic Fixation Processes

1. They require complicated and expensive equipment.
2. Skilled labor is required.

3. They are not applicable for materials that decompose at high temperatures.
4. Oil and odor emissions may occur during heating.
5. The process is energy intensive.
6. Plasticity of matrix-waste mixtures may require that containers be provided for transport and disposal.

Process Application

The thermoplastic fixation processes were developed for use in the disposal of radioactive wastes. This process may be applicable to some hazardous wastes, although many materials may not be thermoplastically fixed. Some of these materials are: (1) organic chemicals that are solvents for the matrix; (2) strongly oxidizing salts such as nitrates, chlorates or perchlorates, which will react with the organic matrix materials and slowly deteriorate the matrix; and (3) salts that will dehydrate naturally at the required mixing temperatures.

Organic Polymer Fixation Processes

The organic polymer fixation process is run as a batch process in which the wet or dry wastes are blended with a prepolymer in a waste receptacle (steel drum) or in a specially designed mixer. When these two components are thoroughly mixed, a catalyst is added and mixing is continued until the catalyst is thoroughly dispersed. Mixing is terminated before the polymer has formed and the resin-waste mixture is transferred to a waste container if necessary [3].

In this process, the waste is not chemically combined with the polymerized material but is trapped in the spongy mass formed by polymerized material.

Advantages of Organic Polymer Fixation Processes

1. High temperatures are not required for forming the resin.
2. The solidified resin is nonflammable.
3. Treated waste does not require drying.
4. There is a high waste:fixative ratio.

Disadvantages of Organic Polymer Fixation Processes

1. The low pH required can put many waste materials into solution.
2. Catalysts may be highly corrosive, requiring special mixing equipment and containment liners.
3. Some cured resins are biodegradable.
4. Materials are trapped, not chemically combined in the matrix.

Process Application

The organic polymer fixation process has been used successfully for disposal of industrial radioactive waste sludges.

Encapsulation Fixation Processes

The encapsulation process involves taking waste that has been bonded previously and enclosing it in a coating or jacket of inert material. The operation of the process requires that the material be mixed with a hardening resin, placed in a mold and heated to produce fusion. The resulting material is a hard, solid block.

Advantages of Encapsulation Fixation Processes

1. The waste material is not brought in contact with the water, enabling highly soluble wastes to be encapsulated.
2. The jacket is impervious, which eliminates leaching.

Disadvantages of Encapsulation Fixation Processes

1. Expensive resins are required.
2. The process is energy intensive.
3. Expensive equipment is required.
4. Skilled labor is required.

Self-Cementing Fixation Processes

The self-cementing process involves treating wastes containing large amounts of calcium sulfate or calcium sulfite so that they become self-cementing. A small volume of waste is dewatered. The sulfite/sulfate sludge is then calcined under controlled conditions to produce a partially dehydrated cementitious calcium sulfate or sulfite. The calcined waste along with proprietary additives are then mixed into the waste sludge. The resulting product is a hard, plaster-like material that is easily handled and relatively impermeable.

Advantages of Self-Cementing Fixation Processes

1. Major additives are available onsite.
2. They are faster setting and more rapid curing than lime-based fixation processes.

3. The final product has good handling characteristics.
4. The final product is stable, nonflammable and nonbiodegradable.
5. The final product effectively retains heavy metals.
6. Waste does not require complete drying before processing.

Disadvantages of Self-Cementing Fixation Processes

1. The process is good only for high-sulfate or high-sulfite sludges.
2. Skilled labor and expensive equipment are required.
3. The calcination process is energy intensive.

Process Application

The self-cementing fixation processes have been used primarily for the fixation of sulfite/sulfate-based sludges produced from SO_2 stack scrubbing operations. However, the process is adaptable for use with calcium sulfite/sulfate sludges produced from other processes.

Glassification Fixation Processes

The glassification fixation process involves mixing waste with silica and fusing the mixture into glass.

Advantages of Glassification Fixation Processes

1. There is excellent containment of wastes.
2. The required additives are fairly inexpensive.

Disadvantages of Glassification Fixation Processes

1. The process is energy intensive.
2. Some wastes may be vaporized during heating.
3. Skilled labor and specialized equipment are required.

Process Application

The glassification fixation process generally is used for the disposal of extremely dangerous or radioactive wastes.

REFERENCES

1. Farb, D. G. "Upgrading Hazardous Waste Disposal Sites Remedial Approaches." U.S. Environmental Protection Publication SW-677, U.S. Government Printing Office, Washington, DC (1978).
2. Fred C. Hart Associates, Inc., Ecology and Environment, Inc. "Remedial Actions for Farm Site, Aurora, Missouri," Unpublished report (1980).
3. Environmental Laboratory, U.S. Army Engineer Waterways Experiment Station Vicksburg, Mississippi. "Survey of Solidification/Stabilization Technology for Hazardous Industrial Wastes," U.S. Environmental Protection Publication 600/2-79-056, U.S. Government Printing Office, Washington, DC (1979).

CHAPTER 7

SITING WASTE MANAGEMENT FACILITIES

Lawrence B. Cahill and Walter R. Holman
Booz, Allen and Hamilton, Inc.
Bethesda, Maryland

For better or worse, United States industry has been, and always will be, generating certain hazardous materials as by-products or waste products of their operations. It is often simply the consequence of supplying the goods society requires. In past decades, ensuring the proper disposal of these waste streams has not been a high-priority environmental issue for industry, government or society as a whole. Emphasis typically has been placed on protection of the air and water environments, the more visual or noticeable waste-receiving media. But now we find that this neglect of the land has led to situations such as Love Canal, which will likely match the harm potential of its water and air counterparts, exemplified by the James River Kepone incident and the continuing Los Angeles smog problem.

In this light, the U.S. Congress, through the regulatory powers of the U.S. Environmental Protection Agency (EPA), has set out to control the indiscriminant disposal of hazardous wastes. Under the Resource Conservation and Recovery Act (RCRA), the EPA has promulgated rules that will help ensure that existing and planned treatment, storage and disposal (TSD) facilities will be designed and perform so as to protect human health and the environment.* Yet the fact is that the many positive effects of these regulations will never be realized if new TSD facilities are not allowed to be sited and constructed in the first place. And indeed there is strong evidence that the siting of TSD facilities will become a key environmental issue of the 1980s. Recent experience has seen some private waste management firms successfully thwarted in their siting attempts by irate local communities, who often view TSD facilities in the same light as nuclear waste dumps. Moreover, in some of these cases the site selected has been near to ideal in most every way: hydrogeologically sound, removed from most residences, designed with safety in mind and financed so that a certain portion of the revenue will be returned to the community in the form of tax rebates. But even here, local officials and residents have fought, and fought successfully, the siting of the facility.

The purpose of this chapter is to examine the siting issue in greater detail and to propose some solutions to the problem. Siting issues can be divided into four categories: technical, economic, institutional and political. Accordingly, the chapter is organized into these sections. As is evidenced from the examples above and the additional discussion below, resolving the issues in total for any given siting attempt likely will be difficult, if not impossible. Yet, *some* sitings *must* be successful if the hazardous waste disposal problem is to see a satisfactory resolution. This chapter is designed to present some ideas on how a solution "package" might be formed so as to encourage the development of a TSD site that will handle the wastes properly and can be accepted by the public.

THE TECHNICAL ISSUES

There are many technical issues associated with the design, siting and ultimate operation of a TSD facility. One of the most important is determining which technology or technologies should be selected. Currently, the predominant technologies in use are land-based disposal and combustion. There is, however, the evidence of a strong push toward more innovative technologies such as plasma wave destruction, ocean incinerators, solidification and ozonation. The principal advantage of these new technologies is that they are typically designed to destroy, convert or recover the hazardous wastes so that none are placed in the ground in their free hazardous state.

*The Resource Conservation and Recovery Act, Section 3004, Treatment, Storage, and Disposal Regulations, 40 CFR Parts 264 and 265, May 19, 1980, and forthcoming.

Typically, selection of a given technology will depend on factors such as the proposed service area, the type of industry(ies) and its (their) associated wastes, the projected size of the TSD facility and the potential for ultimate destruction, conversion or recovery of the hazardous wastes. Experience shows, however, that with respect to siting, it most often makes little difference which technology is selected. Even in cases in which the wastes are completely destroyed by the process (e.g., high-temperature incineration), public opposition has been heavy. In these instances, residents center their concerns first on the transportation of the wastes through the community to the site, prior to destruction and, second, on the possibility, however unlikely, that the process will not achieve full destruction 100% of the time, resulting in occasional hazardous air emissions. Thus, siting success or failure generally has become, insensitive to technology selection.

More to the point with respect to the siting issue is the question one would ask once a technology(ies) and general service area already have been decided on. That is, how do you select a site of several acres from among the total universe of possibilities, whether that universe is an industrial park, a regional industrial service area or an entire state?

Planners recently have been developing methods by which this question can begin to be answered. One proposed solution is to develop a siting criteria framework that will allow the planner to focus in on sites incrementally. As shown in Table I, the framework consists of three levels of siting criteria.*

Level 1 criteria are broad based and generally direct the decision-maker to identify certain critical areas that should be categorically excluded from consideration as TSD sites. Coastal wetlands would be an example of an area where no facility would be appropriate.

Once these broad environmentally sensitive areas are mapped out, the planner would shift to the level II criteria, which are constructed in a more site-specific and positive way. For example, existing industrial zones or public lands that have passed the level I criteria should be identified for further investigation. If there are any locations in the areal universe that could be candidates for sites, the level II criteria should identify them. Typically, and for obvious reasons, industrially zoned land and existing disposal sites (which could be renovated or expanded) are two of the most desirable and suitable areas one could identify.

Finally, the planner shifts to level III criteria, which guide the investigator on the site-specific characteristics that are important for

*This framework has been proposed by Environmental Resources Management, Inc., West Chester, Pennsylvania.

Table I. A Three-Level Siting Criteria System

Level I Criteria
Coastal Flood Hazard Areas
Coastal Wetlands
Public Water Supply Watersheds
Aquifer/Well Yield
Critical Recharge Areas
Seismic Risk Zones
Designated Natural Areas
Level II Criteria
Lands Designated for Industrial Use
Sites of Existing Facilities
Dedicated Lands in Public Trust
Arterial Highways
Level III Criteria
Air (nonattainment area)
Surface Water (flood hazard, flow, quality . . .)
Ground Water (velocity, use, depth, . . .)
Geology (faults, bedrock, . . .)
Soils (permeability, pH, CEC, . . .)
Biological (critical habitat, shellfish, . . .)
Transportation (restrictions, structures along corridor)
Archaeology (historic places)
Socioeconomics (ownership, population density . . .)

identifying the exact location of the facility. Specific criteria range from the natural (air quality, hydrogeology and biology) to manmade (transportation, population density) environments.

The planner would, as a result of the three-level approach, be able to identify several optimal, but not perfect, sites from which a final site could be selected. This final selection process would, of course, involve more than just technical issues. In fact, the issues related to final selection typically become ones of economics, institutions and the political setting.

THE ECONOMIC ISSUES

One of the principal determinants in deciding on particular disposal technologies and, thereby, the type of site needed is economics. Proper disposal of hazardous waste is expensive and there is every reason to believe it will only get more expensive with the advent of the RCRA TSD facility regulations. The central siting issue with respect to economics is whether these high costs may prohibit the ability to develop sites where and when needed.

As shown in Table II, prices and costs for the various technologies vary considerably. Moreover, they are typically very expensive if compared with a benchmark such as nonhazardous solid waste disposal in sanitary landfills, which generally costs $5–$10/ton.

Certain general trends can be observed from the price structure of the technologies. Typically, it costs significantly more to dispose of highly toxic wastes (e.g., PCB). It also costs more to remove any ultimate liability one would incur with a given waste. That is, the technologies designed to destroy the waste (e.g., high-temperature incineration) are more expensive than those that are, in actuality, no more than long-term storage techniques (e.g., secure landfills). Thus, it costs money to have the satisfaction of knowing there will be no nightmares in the future.

It is also well to note that, more than prices, capital investment can be the central economic issue in technology choice. For while industry or government may believe that incineration is the appropriate technology because of its destruction capability, the fact that a median-sized facility may cost $10,000,000 to construct could, in some cases, preclude selection of this option. This could have an unfortunate siting consequence since in some cases destruction technologies are easier to locate, previous statements aside.

Although economics is often a key issue in siting, particularly where local or state governments might be involved in financing all or part of the facility, it is more typically overwhelmed by other issues. And, in fact, much the same can be said of the technical issues as well. Unfortunately,

Table II. The Economics of Hazardous Waste Disposal[a]

Option	Prices ($/metric ton) Standard Wastes	Prices ($/metric ton) Highly Toxic Wastes	Construction Costs ($ for median-sized facility)
Landfill	20–90	100–400	Size dependent
Landfarm	5–25	NA	Size dependent
Incineration	50–300	300–1000	10,000,000
Treatment	15–80	100–500	2,000,000
Deep Well Injection	15–40	50–100	Minimal

[a]Provided by Booz, Allen & Hamilton, Inc., Bethesda, Maryland.

siting decisions often hinge on measures less rational than engineering and economics. Two of these, the institutional and political settings, form the discussion for the remainder of this chapter.

POLITICAL ISSUES

As previously discussed, local opposition to the siting of new (or the continued operation of existing) hazardous waste treatment facilities is related most frequently to a set of political and institutional issues as opposed to the predominantly technical or economic issues. The overview of political issues in this section is organized into three major topic areas:

1. Principal political issues leading to public opposition
2. Characteristics of prior successful siting attempts
3. Tools that can be used to resolve siting conflicts

Within the past five years, much of the public opposition to the siting of hazardous waste management facilities has centered around the siting process itself. The lack of substantive public input during the siting process and the inadequate representation of the public on siting boards has frequently stopped siting attempts in their tracks. This perception of inadequate public representation often gave rise to a wide variety of technical and economic concerns. In many cases, local spokesmen voiced concerns about the use of subjective siting criteria and the apparent availability of superior alternative sites. Furthermore, significant questions were raised concerning the potential economic disbenefits associated with proposed sites and the lack of compensation for these potential disbenefits.

In addition to problems associated with the siting process itself, public opposition has often been triggered or intensified by the perceived lack of credibility of the site sponsor and/or the regulatory agency. The well-documented problems of improper disposal practices experienced by a wide spectrum of the hazardous waste management industry, including the very largest firms in the industry, have eroded the general confidence the public has in this industry. In addition, the flagrant and unsafe practices of a small minority of firms have combined with the public outrage over the Love Canal types of situations to cast a pall over the entire industry. At the state and local level, the lack of adequate resources and the absence of a track record in dealing with such significant issues as hazardous wastes make it extremely difficult for public agencies to gain the public's confidence.

On the other side of the coin, there have been notable successes in siting new hazardous waste management facilities in recent years. Significant insights can be obtained by identifying those factors that differentiate these successful siting attempts from their unsuccessful contemporaries. In this regard, three factors stand apart from all others in terms of importance. First, successful siting attempts are generally characterized by situations in which waste generating industries are crucial to the economic viability of the region. Second, inclusion of public officials and local citizens early on in the site selection process has greatly enhanced the chances for successful siting. Finally, successful siting attempts have been characterized most frequently by situations in which the private waste management firm and the involved public authorities have been held in high esteem by the local community.

In addition to these three factors, successful siting attempts have also been characterized by the exclusion of high visibility wastes, e.g., PCB, and the location of facilities away from major residential areas. Also, these attempts provided a comprehensive demonstration of technical evaluation and planning for proposed sites including the need for the site, adequate safety precautions and explicit assessment of impacts on the local community.

Finally, from the national concern and public opposition to siting of hazardous waste management facilities have arisen several important tools for the promotion of required additional disposal capacity. Among the most prominent are the following:

- state and local government planning in support of new facilities;
- limitation of local zoning;
- mediation of siting disputes;
- provision of incentives for communities to accept sites, and
- public sector participation in the ownership of sites and facilities.

These tools are highlighted briefly in the ensuing paragraphs. The discussion addresses the principal focus of each tool and the precedents for it.

In the case of state and local government planning in support of new facilities, each party plays a lead role in documenting the need for new facilities, in identifying suitable areas for sites, and in establishing the process for site selection. Precedents for this type of assistance are exemplified by the hazardous waste siting boards established during 1980 in Michigan, Minnesota, New York, Connecticut and Massachusetts. In addition, states such as New York, New Jersey and Maryland as well as multistate agencies such as the Deleware River Basin Commission and the New England Regional Commission have undertaken comprehensive hazardous waste inventory/generation studies and preliminary areawide siting studies.

With respect to the limitation of local zoning, several states have enacted legislation that limits the ability of local governments to stop the siting of new hazardous waste management facilities through the exercize of zoning and other land use ordinances. The siting boards in Pennsylvania, Minnesota, Michigan and New York have preemptive authority over local zoning. The primary precedent for this tool derives from the establishment of state authorities in the early 1970s for the siting of electric power plants and major transmission lines.

Another alternative to overcome public opposition to siting entails the mediation of siting disputes. A state or multistate agency would provide, either directly or by contract, an impartial third party to help settle disputes over a proposed facility. This environmental mediation represents an adaptation of techniques used in labor arbitration and intercity conciliation. With respect to precedents in the hazardous waste siting arena, legislation in Massachusetts empowers the state siting board to create an arbitration board if negotiations become deadlocked. Also, the State Office of Dispute Settlement in New Jersey attempted to resolve a dispute over the operation of an existing hazardous waste incinerator in the city of Bridgeport.

The provision of incentives for communities to accept new hazardous waste disposal facilities has attracted much attention within the industry and among federal, state and local governments. The principal focus of this option is the state or multistate state government, which directly provides incentives (or arranges for the private waste management firm to do so) to local communities to compensate for anticipated diseconomies, risks and nuisances associated with proposed sites. Private waste management firms, acting on their own, have offered during siting attempts such inducements as parks, fire equipment and direct financial compensation. At the state level, Connecticut, in addition to giving its siting boards preemptive authority over local zoning, also requires that a new facility pay the local municipality a fee based on the facility's gross receipts. At the federal level, the U.S. Environmental Protection Agency has been assessing the feasibility of a variety of potential incentives including: gross receipts taken on facilities, payments in lieu of taxes, "tipping fees," one-time lump sum payments to residents or local governments, land value guarantees, provision of recreation areas, purchase of buffer zones, and free disposal for the locality or for local business.

The final tool under discussion—public sector participation in the ownership of sites and facilities—is treated in detail in the last section of this chapter. This next section focuses on the primary institutional issues surrounding the siting of hazardous waste management facilities.

INSTITUTIONAL ISSUES

The participation by a state or multistate agency in the ownership of hazardous waste disposal sites and facilities provides a direct way to exert greater public control over the management of hazardous wastes. This type of participation can take a variety of forms, depending on whether the public agency owns the facilities as well as the land, operates those facilities and/or participates in their financing. In general, there are four primary institutional arrangements to facilitate siting:

1. Private ownership and management of facilities
2. State assistance in hazardous waste management and planning
3. Joint state-private sector participation in facility/site ownership and operation
4. Total state control of facility/site ownership and operation

In addition, several states have considered public ownership of the site after closure and the establishment of a postclosure management fund.

In terms of precedents, there are many examples of proposed, attempted and implemented public sector participation of the forms previously delineated. These include: the states of Oregon, Washington, New York, Ohio, California, Arizona and Minnesota; the multistate agencies of the Delaware River Basin Commission and the New England Regional Commission; and the quasipublic state authority—the Gulf Coast Waste Disposal Authority.

With respect to specifics, Table III presents the key features of four alternative institutional arrangements for statewide hazardous waste management under consideration by new York. As demonstrated, these options represent differing degrees of public control and operating/financial risk.

CONCLUSION

We have seen from the discussion that hazardous waste facility siting is indeed a complex issue. Technical issues give rise to questions on selection of the appropriate technology and determining precisely the best technical sites, nontechnical considerations aside. Methods have been proposed that help in these decisions, but the science is not exact. The economics of the situation seems to indicate that already high costs are being driven higher by federal regulations. Moreover, high costs may discourage the use of destruction technologies, which can be somewhat easier to site. In the institutional setting the issues center principally on the involvement of state governments in issue areas ranging from state siting boards with powers to overrule local zoning, to the actual operation of TSD facilities. The jury is

Table III. New York State—Options for State Involvement in Hazardous Waste Facilities

Function	Options		
	Option I Private Ownership	Option II State Assistance	Option III State Ownership Assistance
Siting	Private firms responsible for siting	Active state involvement in siting process: — Stronger state authority in siting — Mediation	Active state involvement in siting process: — Stronger state authority in siting — Mediation
Financing and Ownership			
Site	Private financing and ownership	Private financing and ownership during active life of facility	State financing and ownership during active life of facility
Facilities	Private financing and ownership	Private financing and ownership	Either: State financing and ownership Or: Private operation of facilities
Operation	Private operation of facilities	Private operation of facilities	Either: State operation of facilities Or: Private operation of facilities
Postclosure Management	Private responsibility for postclosure management according to current law	Active state involvement in postclosure management: — Support federal efforts to develop a postclosure liability fund — Develop state-sponsored fund to ensure against liabilities at closed sites — State ownership of facilities after 20 years of closure	Active state involvement in postclosure management: — Support federal efforts to develop a postclosure liability fund — Develop state-sponsored fund to ensure against liabilities at closed sites — State ownership of facilities up to and after closure

still out on the relative merits of state involvement in an area historically left to the private sector. Finally, and perhaps most significantly, the present political climate is such that siting any new TSD facilities without some outside influence over local control of siting decisions may be nearly impossible. With the need for proper disposal options in the country as acute as it is, this is a worrisome situation indeed.

REFERENCES

1. Booz, Allen & Hamilton, Inc. "Options for Establishing Hazardous Waste Management Facilities," report prepared for the New York State Environmental Facilities Corporation, September 1, 1979.

CHAPTER 8

THE SUPERFUND CONCEPT

Michael B. Cook

Associate Assistant Administrator for
Environmental Emergency Response and Prevention
U.S. Environmental Protection Agency
Washington, DC 20460

Legislation setting up a "Superfund" is moving rapidly through Congress at this writing. The legislation would provide funds and authority to the U.S. Environmental Protection Agency (EPA) to control releases of oil and hazardous substances to the environment. The legislation focuses on releases from uncontrolled hazardous waste sites and "spills" or one-time, accidental releases of oil or chemicals.

The Superfund legislation builds on the program currently authorized by Section 311 of the Clean Water Act. The 311 program provides a revolving fund of no more than $35 million to deal with releases of oil or hazardous substances to the surface waters of the United States. The 311 program has been highly successful at responding to spills on an emergency basis. It provides a sound interagency management system consisting of a National Response Team and ten Regional Response Teams. These teams operate under rules established in a National Contingency Plan and have proven to be an effective mechanism for coordinating government action at all levels and reaching agreement on how to handle emergencies resulting from spills.

The National Contingency Plan provides that action at the site of a spill will be directed by an "On-Scene Coordinator." The On-Scene Coordinator has both the responsibility and authority to make decisions on handling emergencies at the scene of the emergency, subject to limitations described in the National Contingency Plan.

The 311 program, although operating effectively within its narrow legislative mandate, has three major shortcomings. First, the 311 funds can be used only for releases of specified substances to surface waters of the United States. It may not be used for releases exclusively to the air, ground or ground water. Second, it authorizes action to stop the release to surface waters but does not authorize major remedial action to eliminate longer-term threats to public health or the environment. Third, the $35 million authorization is pitifully small compared with the costs of remedial actions needed at uncontrolled hazardous waste sites. Replenishment of the funds expended is dependent principally on appropriations of Congress. Some monies also come from recoveries from parties deemed responsible for the releases to the surface waters, where these parties can be found and have sufficient funds to cover costs.

The Superfund legislation would deal with an estimated 10,000–12,000 releases of oil per year and approximately 9000 releases of hazardous substances per year to the environment. Examples of such incidents are oil spilled from an oil tanker after it runs aground, or vapors released from a train wreck with tank cars containing hazardous substances.

The Superfund legislation also would deal with uncontrolled hazardous waste sites found across the entire United States. We currently estimate that there are 30,000–50,000 sites containing hazardous substances. A portion of these sites is currently active and will be subject to permit requirements established under the solid waste program. Another portion of the sites is no longer active and does not represent a threat to public health or the environment because the substances are well contained and are not presently, and likely will not be in the future, releasing substantial amounts of chemicals in a detrimental manner. A third portion of the sites represents some small threat to the public health or the environment, but a threat that is easily managed with resources available to state and local governments. These sites need not receive attention from the federal government and, therefore, would not be handled under the Superfund program.

A final portion of the 30,000–50,000 hazardous waste sites represents a serious problem requiring attention from the federal government. This portion is very roughly estimated at 5000 sites. We believe that roughly this number will require inspection by the federal government or its contractors, and a large number of these sites will require emergency and remedial action. Somewhere between 1200 and 2000 will represent a substantial enough threat to require long-term remedial action roughly estimated now to average $3.0–$3.5 million per site.

The EPA currently has an aggressive program to identify uncontrolled hazardous waste sites and assess the extent of the problem caused by these sites. Where a serious problem is found, emergency action is taken to the extent resources and legal authority permit. Tools available to the Agency include the 311 program, where there is a release to surface waters of a designated hazardous substance. We also can take enforcement action where a responsible party is identified under one or more of several statutes, including the Clean Water Act, the Toxic Substances Control Act and the Resource Recovery and Conservation Act. We also have a very small amount of money authorized during fiscal year 1980 ($1.25 million) to demonstrate new techniques for controlling hazardous substances. This fund has been used at sites where legal authority and monies are not applicable under other programs.

HOW SUPERFUND WOULD WORK

The Superfund would be created by a combination of appropriations from general federal revenue and industry contributions in the form of a fee or a tax. The original administration proposal for legislation called for a fee, but it now seems both preferable and likely that the contribution would be in the form of a tax.

The industrial contribution would come from the oil, petrochemical and inorganic chemicals industries. The tax would be placed specifically on the production of crude oil, petrochemical feedstocks and inorganic raw materials. These feedstocks and raw materials are normally found one step in the refinement process beyond crude oil or beyond the basic ore for inorganic chemicals. The amount of tax on the oil, feedstocks and raw materials would be fixed by law, collected by the Internal Revenue Service, and transferred into a trust fund administered by the Department of Treasury on behalf of the U.S. Environmental Protection Agency.

The Superfund would consist of costs of cleanup and fines recovered from parties found responsible for releasing hazardous substances or oil to the environment.

Pending legislation sets up a fee or tax during a period of five years (for the House bills) or six years (for the Senate bill). The program would have to be renewed or terminated at the end of this period.

The monies in the fund would be used to control releases of oil or hazardous substances on an emergency basis and to take long-term remedial action where appropriate. It might also be used to pay personnel and property damages. The precise extent of the damages eligible will depend on the scope of the final legislation.

The damages covered by the Superfund are among the major political issues being debated during consideration of pending legislation in the House and the Senate. Other issues are the defenses to, and limits of, liability for responsible parties, the size of the fund and the percentage of the fund to be contributed by industry as opposed to general federal tax revenues. These issues are discussed in more detail in the following sections.

SIZE OF THE FUND

The pending Superfund bills all provide about the same amount of money per year for emergency containment and long-term remedial action to cope with releases of oil and hazardous substances to the environment. Both the House and the Senate provide for about $2 billion over five years for control of releases. The Senate also provides additional funds for a sixth year and for extensive damage claims provided in their version of the Superfund legislation.

The size of the fund remains a political issue because the House Interstate and Foreign Commerce Committee originally provided a substantially smaller amount of money for cleanup of sites than is provided in the final bill reported out of the House Ways and Means Committee. The differences between the two committees suggest the possibility of action on the floor of the House of Representatives to reduce the size of the fund.

INDUSTRY VERSUS FEDERAL CONTRIBUTIONS TO THE FUND

The pending Senate bill would require industry to contribute 87.5% of the money in the fund, and 12.5% would come from general federal revenues. Pending House bills would require a slightly smaller percentage contribution from industry and a commensurately larger contribution from general revenue.

The relative contributions of the government and industry to the fund remains a political issue. The chemical industries believe quite strongly that the federal contribution to the fund should be a much larger percentage. They also believe that expenditure should be allowed from the fund only to the extent that the Congress appropriates revenues in the proportion provided under legislation. If the Congress does not appropriate fully the amount authorized from general revenues, then the funds collected from industry in the form of taxes or fees could not be utilized in full and would have to be set aside to the extent that the authorized matching funds were not appropriated.

THIRD PARTY DAMAGE CLAIMS AGAINST THE FUND

The pending legislation in the House authorizes damage claims against the fund for property, natural resources and loss of opportunity to harvest marine life. The Senate bill goes farther and authorizes damage claims against the fund for lost income due to property damage or personal injury, out-of-pocket medical expenses, rehabilitation and burial expenses, epidemiological studies and diagnostic services, and loss of tax revenues for one year.

These third party claims against the fund could conceivably absorb huge sums of money. Authorizing such claims is attractive because of the very serious economic and health impacts of some releases to the environment. The principal drawback for such claims is that they would be difficult to limit to the most seriously impacted victims. The state-of-the-art linking particular health problems to known exposure to hazardous chemicals at uncontrolled waste sites is embryonic, at best, and does not lend itself well to distinguishing the validity of claims among persons living near uncontrolled waste sites who are somehow exposed to releases from these sites.

LEGAL LIABILITY PROVISIONS

The chemical industry has strongly resisted tough liability provisions in the Superfund legislation. The administration, on the other hand, has sought strict, joint and several liability, which would mean that any party found responsible for a release of hazardous substances could be held liable for the full costs of remedial actions and eligible damage payments associated with the release. Defenses to liability would be extremely limited and there would be no third party defenses explicitly provided in the law. Courts would, of course, have the discretion to apportion costs among responsible parties, but such apportionment would not be mandatory.

The administration believes that common law, in dealing with the kinds of damages covered by Superfund, is in the process of articulating a position very close to strict, joint and several liability. This position has been largely adopted by the Senate and by the portion of the House action to date dealing with spills. The House proposal on sites, however, would depart substantially from strict, joint and several liability by providing third party and other defenses and imposing a mandatory apportionment of costs among responsible parties. Such limited liability would make recovery of costs to the fund difficult and slow, if litigated.

CONGRESSIONAL COMMITTEE JURISDICTION

Superfund legislation ranges broadly and has been receiving attention in several committees of Congress. The House Merchant Marine and Fisheries Committee reported out a bill covering only oil spills. This bill was referred to the Public Works and Transportation Committee, which added a title on chemical spills. The House Interstate and Foreign Commerce Committee reported out a bill covering uncontrolled hazardous waste sites. The House Ways and Means Committee then took jurisdiction over the revenue provisions of both the spills bill and the sites bill, and reported out revenue provisions that varied substantially from those provided in these two bills.

The Senate Environment and Public Works Committee has reported out a bill covering spills of hazardous substances and waste sites. It does not cover oil spills, although such provision may be added on the floor of the Senate. The Senate Finance Committee has taken jurisdiction of the revenue questions in the bill and is expected to act on these shortly.

The intense public and congressional interest in Superfund and the progress already enjoyed by Superfund legislation in the Congress makes a new law likely at some time during the near future. The timing and exact form of the legislation will depend, to a large degree, on the ease of resolving the major issues discussed above and the priority afforded to legislation in the Houses of Congress over the next several months.

SUPERFUND IMPLEMENTATION PLANNING

The EPA has established a small office to prepare the federal government for implementation of Superfund. The Office of Analysis and Program Development (OAPD) is located in the Office of Water and Waste Management with the 311 spills program. The OAPD presides over seven task forces set up within EPA to analyze major issues and take the plethora of actions necessary to implement what will be one of the largest programs administered by the Agency. Regulations, guidelines and procedures are being drafted; managerial systems are being developed;

automated data processing is being designed; and a new organizational structure is being developed for Superfund.

The Agency is also working closely with other federal agencies on the National Response Team to prepare a preliminary draft revision to the National Contingency Plan. The revised plan would provide for implementation of the Superfund.

The Agency is also working closely with the National Governors Association to discuss the future role of the states in the Superfund program. The pending legislation would allow EPA to contract with states, other governmental entities and private parties to administer portions of the Superfund program. The Agency assumes that it will have to depend, to some extent, on these other parties for Superfund administration because of overall controls on the federal budget and manpower.

The Agency is preparing an extensive program to foster and facilitate very extensive participation by public, private and special interest groups in Superfund implementation. This program will be implemented immediately after enactment of the Superfund bill. Of course, the Agency will have to work very closely with the engineering profession in implementing the wide-ranging program provided under Superfund.

CONCLUSION

Superfund builds on a highly successful emergency response program authorized under the Clean Water Act. It provides the federal government with authorization and the money needed to deal with widespread, serious problems created by uncontrolled hazardous waste sites, such as Love Canal, Valley of the Drums, Chemical Control in Elizabeth, New Jersey, and Stringfellow Disposal Site in California. Enactment of the legislation is likely in the near future if several key political issues can be resolved. Meanwhile, the EPA is preparing to implement the new legislation. The public perceives the uncontrolled waste site problem as extremely serious and immediately threatening the health and well-being of many Americans. Implementation of the new legislation will have to proceed very quickly to respond to the interests and concern being displayed by the general public and the Congress. The Agency will be ready to proceed in close partnership with federal, state and local governments and the interested private sector, including the engineering profession.

PART 3

CASE STUDIES IN THE MANAGEMENT OF HAZARDOUS WASTE

CHAPTER 9

THE KEPONE CLEANUP

John D. Steele

Drake Engineering Company
Richmond, Virginia

William F. Gilley

Director, Division of Solid & Hazardous Waste Management
Virginia Department of Health
Richmond, Virginia

The pesticide Kepone was first developed by Allied Chemical Corporation, Morristown, New Jersey during the late 1940s and early 1950s. The compound was then patented in 1951.

Kepone is manufactured by reacting hexachlorocyclopentadiene (HCP), sulfur trioxide (sulfan), sodium hydroxide (caustic), sulfuric acid and antimony pentachloride.

In March of 1966, the production of Kepone was consolidated at Allied Chemical's Semi-Works plant in Hopewell, Virginia. Operations continued there until 1974, when Life Science Products went into operation.

Life Science Products Company (LSPC) was established for the sole purpose of manufacturing Kepone as a toll operation for Allied Chemical, and it is apparent that from the start the fate of the business rested with LSPC's agreement with Allied Chemical.

This company started production during the last week of February 1974. In February 1974 there was no government organizational structure within the Commonwealth capable of responding to a widespread or severe environmental crisis. Virginia's plan for the administration of the Occupational Safety and Health Act (OSHA) had been submitted to the federal government for approval but was not yet effective; therefore, Virginia laws on this subject were preempted by OSHA. In addition, there were no state laws on the books specifically directed toward the regulation of the manufacture and distribution of toxic substances.

Almost immediately after the startup of the LSPC in February 1974, a long chain of events occurred that led to the closing of the plant in July 1975. On March 6, 1974, the State Air Pollution Control Board cited LSPC for failure to apply for a permit and for excess leakage of sulfur trioxide. LSPC promptly complied with the statutes by submitting the required information, and the permit was approved on July 19, 1974.

The information on the forms identified Kepone as the product being manufactured at LSPC but did not reflect that it was in any way hazardous.

In March 1974 the Hopewell Sewage Treatment Plant anaerobic sludge digesters began to go "sour." The digesters failed one month later, resulting in the landfilling of the sour sludges in the adjacent city landfill.

In the resulting meetings with state and local officials, it was felt that the upset at the Hopewell Sewage Treatment Plant could have been caused by Kepone. In addition, during these sessions it was decided that the sludges existing in the digesters should be pumped into an especially constructed asphalt-lined pit. After authorization from the State Health Department on December 10, 1974, the project was begun.

On May 28, 1975, a turning point occurred in the events. Sometime during the day a worker at the Hopewell Sewage Treatment Plant was overcome by fumes from hexachlorocyclopentadiene. Fortunately, the worker was not seriously affected. The incident was reported to the State Water Control Board and the State Health Department. Although the effects of Kepone may have been relatively unknown at this time to the state regulatory personnel, the ingredient HCP was a fairly commonly known chemical, the toxic effects of which were well known to the department's industrial hygiene personnel. On the next day, an inspection was made by Health Department personnel during which relatively high levels of HCP were identified at various locations within the sewage treatment plant. The Williams-Steiger Act of 1971 gives exclusive authority for uninvited inspections to OSHA; consequently, State Health Department personnel were required to wait until they were invited to the plant site. Due to various delays, the actual inspection was not conducted until July 22, 1975. Because of the findings, an order for closure of the plant was prepared by the Attorney General's office.

On July 24, 1975, officials of the state met with the owners of LSPC. The owners agreed to comply voluntarily with the conditions of the order and so the order was not officially served. The next day cleanup operations were initiated and the process of the demolition and removal of the LSPC process had begun.

TASK FORCE ORGANIZATION

The responsibility and authority for the environmental protection in Virginia was, at the time of this incident and is now, split between the State Water Control Board and the State Health Department. In 1975 there was no state agency specifically dedicated to dealing with hazardous waste. Because of the medical implications and their early involvement, the State Health Department assumed the lead role in initial investigations dealing with Kepone contamination.

After the rough assessment that both human and environmental contamination were severe and extensive in the Hopewell area, the Governor of the Commonwealth created a task force in the winter of 1975 to deal with the crisis. This task force was composed of members from:

- Department of Health
- State Air Pollution Control Board
- State Water Control Board
- Office of Attorney General
- Division of Consolidated Laboratory Services
- Department of Labor and Industry
- Department of Agriculture and Commerce
- Virginia Commonwealth University, Health Sciences Division
- Virginia Institute of Marine Sciences
- Virginia Polytechnic Institute and State University
- Virginia Marine Resources Commission
- Health, Law and the Environment, Office of Emergency Services
- City of Hopewell, Ex-officio
- Environmental Protection Agency (EPA) (Coordinating)
- U.S. Federal Food & Drug Administration (FDA) (Coordinating)

Various subcommittees were established to carry out specific tasks. Among these were:

- Committee for Sludge Disposal
- Committee for Dredging
- Committee for Marine & James River Fisheries
- Committee for Laboratory Resources & Analytical Procedures
- Committee for Community Environment

With an official status and the authority to carry out its previously unstructured plans, the task force, through its subcommittees, established as first priorities determining the extent of contamination and ascertaining a methodology for disposing of the waste.

At this point, the Kepone crisis had received wide national media coverage resulting in emotional reaction that sometimes resembled a fictitious horror story. It became immensely difficult for the task force to proceed with its work and maintain a clinical, logical approach to the problem. Unquestionably, the Freedom of Information Act was not designed for conditions such as these.

EXTENT OF CONTAMINATION

While the medical problems were being assessed, an extensive sampling program was begun to determine the extent of terrestrial, water and air contamination. As the degree and extent of the contamination became more defined, the materials fell into the following categories:

- Cleanup waste from demolition of the LSPC manufacturing process, which included building materials, paving, washed-down water from manufacturing equipment such as reactors, tanks and pumps; piping, clothing and other miscellaneous waste.
- Contaminated soil in the vicinity of LSPC, which represented several tons of material that had been excavated.
- Sewage sludges from the Hopewell Sewage Plant, including over 1,000,000 gallons of sludge held in a lagoon, plus an undetermined amount of sludge that had been discharged into the landfill immediately adjacent to the sewage treatment plant.
- Contaminated soil from various dump sites in the Hopewell area where LSPC had disposed of off-standard batches of Kepone.
- Demolition waste from the original semi-works plant located on the Allied Chemical complex.
- Contaminated swamp material from the Bailey's Creek swamp area immediately below the Hopewell landfill, which exceeded 400 tons of drummed material.
- Miscellaneous areas in the landfill that were used as dump sites for Kepone by LSPC.
- The portion of the James River extending from the vicinity of Hopewell to the mouth of the Chesapeake Bay, which involves approximately 65 miles of river and approximately 60,000,000 square yards of river bottom. It is estimated that the river contains approximately 40,000 pounds of Kepone.
- Twenty-five tons of commercial-grade Kepone stored on U.S. Navy property in Portsmouth, Virginia, plus 330 tons of Kepone and Kepone-contaminated material stored in Baltimore, Maryland.
- Twenty rail cars of contaminated water from various washed-down operations stored on Allied Chemical property.

This assessment to determine the degree of contamination took from the winter of 1975 through mid-1977 and there are areas still under examination. Currently, monitoring continues in the James River, Chesapeake Bay, Hopewell landfill sites and disposal areas.

During the time frame when the task force was investigating the extent of contamination, EPA, in association with FDA and the National Cancer Institute (NCI), was performing hasty studies to determine an action level for the biota in the James River. Eventually, FDA established action levels of 0.3 ppm for finfish and 0.4 ppm for crabs. Those action levels remain in effect today and have affected the fisheries industry in the James River from the time of its closure in December 1975.

DISPOSAL ALTERNATIVES

From late 1975 through 1978 various methods of disposal were investigated by numerous individuals and organizations. At the onset, one of the first problems was the lack of knowledge about the compound. Although Allied Chemical's technical and scientific groups showed a willingness to share some information concerning the compound, legal considerations precluded an open exchange of information. Much was known about the manufacturing process; however, very little was known by anyone about potential destruction methods. The problem of disposal was further complicated by the fact that the Kepone was bound up in various forms such as construction materials, sewage sludges, water and river sediments. Virginia's first priority for disposal of contaminated material was destruction; therefore, a search for the appropriate method was initiated in late 1975.

Thermal destruction is a common method of disposing of pesticides, and though much was known about the subject there was only one experiment known that dealt directly with thermal destruction of Kepone. This procedure was described by Duvall and Rubey at the University of Dayton Research Institute (UDRI), who performed an experiment known as "Laboratory Evaluation of High-Temperature Destruction of Kepone and Related Pesticides." This work was conducted in the early spring of 1976 under an EPA grant and reported in May 1976. The experiment showed that Kepone and its degradation by-products, which were largely chlorinated hydrocarbons, were efficiently destroyed at 1000° C (1930° F) at a residence time exceeding 1 second when heated in dry air.

Although this experiment was a step in the right direction, it by no means guaranteed the success of a destruction process under flame conditions using state-of-the-art equipment for large-scale incineration and using Kepone combined with many other materials. So the Commonwealth, in conjunction with EPA, developed an experimental protocol for a project called KIT—Kepone Incineration Test.

The development and approval of the project was in itself a major undertaking since by this time public information concerning the compound had been blown out of proportion to the extent that no one anywhere wanted to have even the slightest contact with Kepone. Eventually the project required the approval of 42 federal and state agencies, which alone was a substantial accomplishment. The project was coordinated by the consulting engineering firm of Design Partnership from Richmond, Virginia, with analytical and scientific assistance from Versar, Inc. of Springfield, Virginia. The actual incineration was performed at the Midland-Ross Industrial facilities in Toledo, Ohio. The test was a success as documented in EPA's Report 600-2-78-108 entitled "Kepone Incineration Test Program." The key technical recommendation was that the destruction efficiencies may exceed 99.9999% if the system consisted of a fume incinerator capable of sustaining 1000° C with sufficient volume to allow a two-second residence time, a pyrolyzer that could continually operate at 450° C, which vaporized and partially decomposed the Kepone; and a scrubber system using a solution with a pH above 9.0. The entire system must operate at a negative pressure and be fitted with the appropriate controls and sensors to maintain operational conditions.

The conditions under which this test was conducted represent a facinating story on its own merits. For instance, the stack emissions limit imposed by the group conducting the test on themselves was 1 mg/m^3 in the workers' environment. Other considerations that were extremely important in this project were the method used to establish the organization and management structure; the health and safety requirements; analytical procedures; and weather conditions encountered during the test.

Since the incineration test provided the base data for the fabrication of a large-scale incineration unit, the next task was to develop a Facility Plan that would incorporate such a unit as an alternative for the long-term disposal of Kepone and Kepone-contaminated waste. The Facility Plan was developed in the same format as the standard EPA 201 Facility Plan. Through conversation with EPA it was determined that Construction Grant Funds could be made available for such a project. Therefore, the procedure for developing a plan and the format of the document were in accordance with the EPA regulations. It was anticipated that if the Facility Plan was approved by the Commonwealth and accepted by the public, then federal funding would be sought.

The planning process addressed nine disposal alternatives as follows:

1. Do nothing
2. Biodegradation
3. Incineration at sea

4. Land-based incineration
5. Long-term containment
6. Modification of Hopewell Regional Sewage Plant Incinerator
7. Commercial incineration facility
8. Catalytic reduction
9. Cement kiln incinerator

Each of these alternatives was evaluated on the basis of public health and safety, monetary cost, environmental impact, public acceptability, implementation, occupational health and safety, reliability and future benefits. As a result of the screening process involved in the plan, three alternatives were selected for further investigation:

1. incineration at sea
2. land-based incineration
3. commercial incineration facility

In further considering each of these alternatives, cost estimates were prepared, preliminary layouts developed and engineering investigations performed that included design and cost information from manufacturers of applicable equipment. This work was completed during the fourth quarter of 1977. As part of the planning process, public hearings were held in the Hopewell area to receive input from interested parties. During these meetings it was apparent that public sentiment strongly opposed the installation of an incineration facility in the Hopewell area and that the public much preferred the alternative that involved incineration at sea. Because of public opinion and other conditions it was decided to delay a final decision on the incineration program until other emerging possibilities were investigated more fully.

JAMES RIVER STUDIES

Although the land-based Kepone and Kepone-contaminated waste imposed the most immediate threat to health, the contamination problem with the James River was under continual study. As indicated previously, the immediate solution to keeping the river from becoming a health hazard was to close the river to fishing and the harvesting of crabs and oysters. At the onset of the identification of contamination, the Virginia Institute of Marine Science (VIMS) and EPA developed and began a series of studies on the river. The task force objectives with respect to the James River studies were to determine:

1. a "no effect" level of Kepone in the river;
2. technical procedures that could economically reduce Kepone to a "no effect" level in the river; and
3. a method for the collection and disposal of contaminated spoils from the James River.

The VIMS studies focused on the effects of Kepone in the biota and the mechanisms involved in the transport of Kepone both vertically and horizontally within the river system. The biological studies examine the amount of Kepone available in the food chain, water and sediment. They also develop information on the effects of Kepone on the growth, reproduction and mortality of various species in the river system. Although many types of shellfish, fish and plants were studied, particular attention was paid to organisms that were of the highest economic consequence. In addition to the biological work, VIMS developed models and other predictive tools to determine the movement of Kepone within the river. Although much of this work has been completed, efforts on the models and some biological studies are yet to be finished. In the final analysis, these studies are directed toward determining what must be done to reduce the reservoir of Kepone to a point at which the levels in the biota are below the FDA action level of 0.3 ppm for finfish and 0.4 ppm for crabs.

The EPA report—"Mitigation Feasibility for the Kepone-Contaminated Hopewell/James River Areas"—is a document that encompasses extensive studies performed through other federal agencies. The report includes appendixes developed by the U.S. Department of Energy, Battelle Pacific Northwest Laboratories; Department of the Army, Corps of Engineers, Norfolk District; and EPA Gulf Breeze Environmental Research Laboratory and Virginia Institute of Marine Science. The EPA summary discusses the background of the contamination problem, the distribution of Kepone and magnitude of contamination, and examines various mitigating techniques. The Corps of Engineers work explores the methods for containing Kepone in the Hopewell area, including the tributaries of the James River and removing contaminated sediments from the James River and its tributaries. The Corps report quantifies the Kepone in affected areas and shows alternatives for structurally containing the material, including cost estimates and schedules for completing various removal or containment projects.

The Battelle work concentrates on the land-based contamination issue and addresses such concerns as the ecological effects and disposal methods of Kepone and Kepone-contaminated material. Battelle employed extensive field sampling in developing its report and looked into unconventional methods of disposal.

The Gulf Breeze study contains a series of reports that parallel and partially overlap the VIMS work. These studies deal with toxicity in marine organisms, bioconcentration and aquatic toxicology. The organisms examined ranged from algae to crabs and included various varieties of minnows and finfish.

ADDITIONAL STUDIES

As indicated in the discussion concerning the Facility Plan, various disposal alternatives were investigated as part of the planning process and numerous nonconventional methods were examined as separate projects. The incineration at sea alternative involved two ships that had been used in Europe for the destruction of toxic compounds, but it was decided that due to international problems and extensive modifications required for pollution control devices the alternative would be abandoned. The multiple-hearth incinerator at the Hopewell Regional Sewage Plant proved to be far too expensive to modify for the purpose of Kepone incineration. Catalytic reduction represented an untried process that was time consuming and extremely expensive to operate.

Biodegradation remains a possibility for disposal. Preliminary experiments indicated that Kepone was biodegraded through anaerobic digestion of sewage sludges. This phenomenon was supported by one bench-scale experiment, but in a subsequent experiment the process did not work effectively. Originally, Kepone in the digesters was reduced by approximately 50% in a 45-day time frame.

Various other methods and types of hardware were examined such as liquid injection process, fluidized bed process, molten salt process, wet oxidation process, plasma destruction, photochemical degradation, radiation and catalytic reduction. Information concerning the advancement of these methods and possible application to Kepone and other pesticides is being monitored continually.

DISPOSAL METHOD

In early 1978 Allied Chemical, through extensive negotiations, had reached a tentative agreement with an organization known as Kali and Salz of Kassell, West Germany and with the Bad Hersfeld Mining Authority to dispose of up to 350 metric tons of Kepone and Kepone-contaminated material in an abandoned salt mine in West Germany. This action, which resulted in a contract completed during July 1978, provided a definite disposal method for the bulk of commercial-grade Kepone stockpiles and the highly contaminated material. In exercising this option, Allied planned to ship the entire stockpile of Kepone residues in Baltimore, Maryland, the drums in the Portsmouth Naval Facility in Portsmouth, Virginia, and the highly contaminated drummed material from the storage site at the Hopewell Sewage Treatment Plant. After a series of conferences and agreements, the Commonwealth gave conditional approval to ship the material. On December 1, 1978 this shipment took place. The logistics of the shipment were relatively simple: the drums from the Hopewell Sewage Treatment Plant were loaded onto trucks contracted by Allied Chemical and driven to the commercial docks in Portsmouth, Virginia, where they were loaded onto the ship. Similarly, the drums at the Naval base were trucked to the dock for shipment. There were no problems encountered during the transportation or loading process.

As part of the negotiations with Allied Chemical concerning the shipment of drums to Germany, Allied agreed to construct a containment cell for the sludges in the lagoon at the Hopewell Treatment Plant site. Prior to this time, Allied had constructed a similar cell on its property in Hopewell for the disposal of low-level waste, which was a result of the demolition of the semi-works plant that had produced Kepone prior to Allied's contract with LSPC. In addition to the sludges, the cell would contain drums from the cleanup of the swamp below the Hopewell landfill and other low-level contaminated material. The cell was constructed at the site of the Kepone lagoon. The cell designed required the installation of a stone base, underdrain system, minimum of three feet of compacted clay, synthetic liner, gas collection system, security system and continuous monitoring system to detect any leaks. The project was complicated by the necessity to remove and protect the contaminated sludges during the construction of the cell. In addition, the excavated area had to be kept dry to keep the project on schedule. Although there were a multitude of construction problems during the project, the cell was completed in the late fall of 1979.

With the completion of the containment of the known quantities of terrestrial Kepone, efforts were intensified to resolve issues in the James River. At this time all but a few of the biological reports on the James River are in final draft form. Those reports indicate that Kepone has no substantial effect on the mortality and ability to reproduce of the species studied. Efforts continue on calibrating the models, and computer runs on the James River Response Model were to be made using the data from the biological studies before the end of 1980. The results of runs from the Response Model and the Transport Model are essential in predicting "no effects" conditions in the James River. Parallel with the VIMS work on the James River are efforts by the Commonwealth to more fully understand the health effects of Kepone. The Health Department has involved such institutions as the Medical College of Virginia and Johns Hopkins University, as well as the number of well-known toxicologists in monitoring the human health effects of Kepone. Studies such as the depuration experiments performed at the Medical College of Virginia have yielded results promoting optimism that the long-term effects of the pesticide may not be as dangerous as previously thought. Hopefully the combined river studies and health effects work will continue to indicate that Kepone's effects on humans is reducing and that perhaps a reexamination of the action level would be in order. The Commonwealth is seeking every opportunity to return the full utilization of the resources of the James River to the public, even if it is only possible to allow commercial fishing of selected species during restricted time periods.

CHAPTER 10

OHIO RIVER PARK, ALLEGHENY COUNTY, PENNSYLVANIA: A CASE HISTORY OF LOCAL GOVERNMENT INVOLVEMENT IN THE ASSESSMENT OF AN ABANDONED HAZARDOUS WASTE DISPOSAL SITE

Albert H. Brunwasser
Environmental Health Services

Paul L. Spence
Allegheny County Health Department
Pittsburgh, Pennsylvania

In 1976 a 35-acre site at the western tip of Neville Island in the Ohio River, Pennsylvania was donated to Allegheny County for the purpose of developing a recreation park. The Allegheny County Department of Planning and Development was responsible for the overall management of planning and construction for the site's conversion to a \$3.3 million recreation facility. During the development of Ohio River Park, various industrial residues and municipal wastes were encountered at and near the ground surface. The Allegheny County Health Department (ACHD) became involved in this matter in July 1978 when it was asked by the County Planning Department to investigate fumes emanating from certain areas of the park where buried waste material had been disturbed by construction activities. ACHD found that little reliable information was available on the types and quantities of wastes deposited at the site.

Therefore, ACHD directed the County Planning Department to obtain the services of a consulting firm to investigate these issues and to assess potential health implications. ACHD found this report inconclusive and was subsequently charged by the Allegheny County Board of Commissioners with investigating the potential public health hazards.

In response to this request, ACHD conducted a search for an engineering consultant experienced in hazardous waste site investigations and selected Fred C. Hart Associates, Inc. (FCHA) of New York. FCHA was charged with determining whether a public health hazard existed at Ohio River Park. ACHD worked closely with FCHA on every phase of the investigation, which was performed during June and July of 1979.

During the investigation, historical background information was gathered and interpreted. Surface wastes were mapped by visual inspection and by delineating the likely locations of buried waste by application of electrical resistivity and metal detection survey techniques. Solid, liquid and gaseous wastes from the surface and near surface were sampled and analyzed. Finally, the toxicological aspects of the waste sampled were assessed and possible exposure mechanisms identified.

Numerous toxic chemicals such as benzene, phenols, parathion and coal tar residues were detected in the wide range of samples collected. Considering the chemicals present at the park site, their observed prevalence, measured concentrations, toxicological and synergistic properties, and the possible exposure mechanism of park visitors to those chemicals, users of the park could experience adverse health effects. Thus it was concluded that a public health threat existed at the site.

As a result of the Phase I assessment, FCHA was commissioned to undertake a second study in the fall of 1979. This study was designed to determine potential offsite environmental effects resulting from the release of contaminants from the park, as well as to delineate and assess alternative remedial measures for the site. Once again, the Phase II study involved a close working relationship between ACHD and FCHA. It specifically included:

1. additional field inspections of site;
2. additional metal detection and electrical resistivity surveys;
3. in-depth interviews of park construction personnel;
4. installation of six ground water monitoring wells for ACHD and two additional wells for the U.S. Environmental Protection Agency (EPA);
5. boring logs and grain-size analysis of soil samples taken during test well installations;
6. sampling of ground water from test wells and sampling by ACHD of surface water and storm water discharge; and
7. analysis of water samples for an array of toxic and related constituents.

Utilizing data from these sources, it was found that the ground water was contaminated and that small quantities of toxic pollutants were entering the Ohio River through surface runoff and ground water seepage. Because this pollution has been adding to the overall degradation of the river water potability, periodic organic monitoring and analysis of downstream water supply intakes was recommended.

Possible mitigative measures were considered to eliminate or reduce the impact of the hazardous waste at the park and were presented to Allegheny County in January 1980. The remedial options ranged from abandoning the park and placing an impervious clay cap over the site at an estimated cost of $250,000, to removing all contaminated waste from the site at an estimated cost of $7–$24,000,000.

During this two-phased investigation, Allegheny County expended over $100,000 in county funds for work performed by the private consultant. Moreover, ACHD developed the scope of work that FCHA followed and closely monitored FCHA's work throughout the entire study. It should be noted that the EPA and the Pennsylvania Department of Environmental Resources (PennDER) provided various levels of technical assistance during this undertaking.

At this time, Allegheny County has negotiated the return of the park site to the former owner, who originally donated the land to the county. The county has been reimbursed for the funds it expended developing the park and investigating the park's hazards. The present owner, the Hillman Foundation, has stated it plans to hire another private engineering consulting firm to both monitor and reassess the situation at the park. In the meantime, ACHD is maintaining a close watch on downstream water supplies while additional studies and courses of action on this matter are being considered.

BACKGROUND

Neville Island is located in the Ohio River approximately 5 miles from the Golden Triangle of Pittsburgh, Pennsylvania (Figure 1). The Golden Triangle is formed by the confluence of the Allegheny and Monongahela Rivers, which form the Ohio. Along these rivers exist some of the most heavily industrialized areas of the world. This is especially true along the Monongahela, where there is a heavy concentration of primary steel manufacturing plants.

The island is 6.5 miles long and only 0.75 of a mile wide at its widest part. The total land area is 868 acres. To the north of the island is the deep,

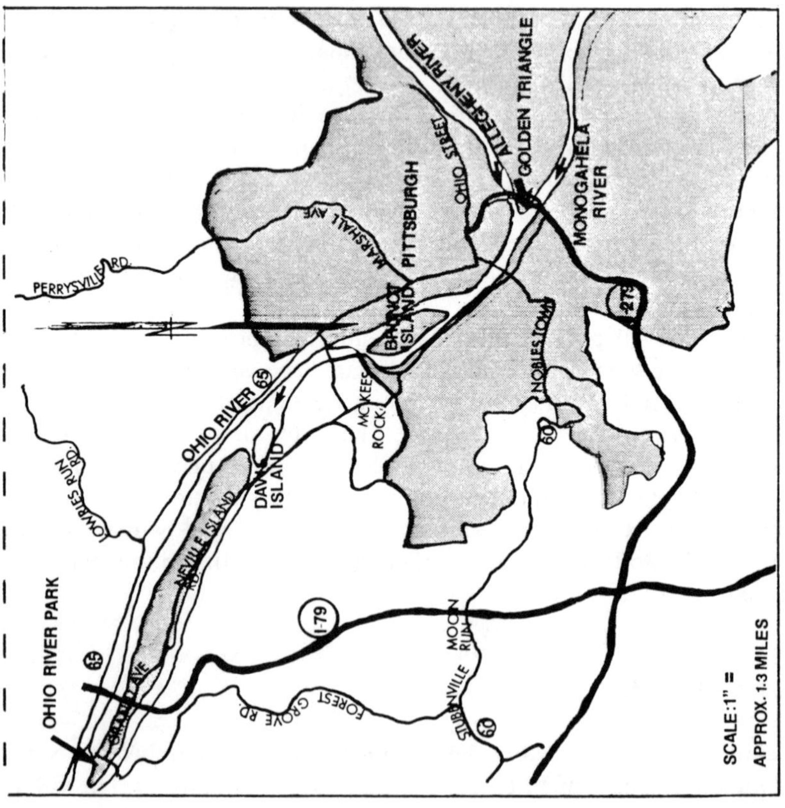

Figure 1. Vicinity map.

heavily traveled main channel of the Ohio River; to the south, a mud-banked untraveled back channel.

Geologically, Neville Island was formed on an alluvial deposit consisting of layered sands, silts and clays. The deeper buried valley deposits under the island consist mostly of sands. Together with the high permeability of these soils and the proximity of the Ohio River for recharge, these buried valley deposits are a valuable source of ground water to the residents of Neville Island and the vicinity.

Because of its location in the Ohio River with its fertile soils and abundant water supply, Neville Island was once called the "Gem of the Ohio" for its agricultural prominence. In the 1800s, farming was dominant and asparagus grown on Neville Island even reached the tables of some of the fine East Coast restaurants. Also, sod for the Pittsburgh Pirates' former home ballpark of Forbes Field supposedly was obtained from Neville Island during the early 1900s.

However, during the turn of the century, the United States was quickly becoming a major industrial power and the Pittsburgh area certainly did its part to bring this about. Neville Island had two things that these new industries needed:

1. water (surface and ground) for their manufacturing processes and energy supplies; and
2. a river system on which to transport their raw materials and products.

Thus began a transition of an agricultural river island to a diversified industrial community.

Prior to World War I, there were 9 industrial plants located on the island. By World War II, there were 23 major industries, employing one-half of the island's population. In the early 1950s the dominant industries were chemical and steel fabricating plants (Table I). One of the chemical plants was operated by Pittsburgh Coke & Chemical, which produced coke, pig iron, cement and agricultural chemicals, while the other was run by the Neville Company, which manufactured resins and solvents, as well as various chemicals. The products manufactured by the other companies were extremely varied, ranging from gasoline and kerosene produced at Gulf Oil's refinery to tin and scrap reclaimed by Vulcan Detinning. Of course, this diversity of products manufactured resulted in a wide range of by-products or waste, which had to be disposed of in some manner.

As was the case in many industries in operation in the 1940s and 1950s, some of the island's industries discharged their industrial waste effluents directly into a watercourse like the Ohio River, while others used the sanitary sewer system. Another waste management option frequently exercised by some plants was the disposal of waste by-products onsite,

Table I. Major Industries Located on Neville Island in 1950

Air Reduction Company	Neville Company
American Tubular Elevator Company	Neville Concrete Pipe Company
Cementstone Company	Neville Island Glass Company
Concrete Products Company	Pittsburgh Coke & Chemical Company
Dravo Corporation	Pittsburgh-Des Moines Steel Company
Foundation Company	Pittsburgh Screw & Bolt Company
Frick & Lindsay	Shenango-Penn Mould Company
Gulf Oil Corporation	Vulcan Detinning Company
Keystone Solvents Company	Watson Standard Company
Lee C. Moore Company	Waverly Oil Works Company
Mallinger Construction Company	Pittsburgh Barrel & Drum Company
Neville Coke Company	

either in lagoons and pits or by directly applying it to the land. However, some facilities had to use offsite land to dispose of their unwanted industrial residues.

During this period, one such offsite area existed at the downstream end of Neville Island. This undeveloped, flood-prone property was owned by Pittsburgh Coke & Chemical, an industrial resident of the island. Four acres of the site allegedly were operated as a garbage dump from 1935 to 1945. In the early 1950s, large quantities of miscellaneous industrial wastes reportedly were deposited extensively in the area.

It is not clear which companies, both on or off the island, used the Pittsburgh Coke & Chemical property as a disposal site. However, due to the nature of their product, lack of sufficient area for captive disposal sites and their close geographical proximity to this vacant property, it is reasonable to suspect that industries located on the island used it as a dump. Some possible wastes generated by these industries between 1940 and 1960 and subsequently disposed were:

- coke distillate by-products
- off-specification pesticides, herbicides, insecticides
- off-specification pigments for paints and dye-stuffs
- still bottoms
- settling basin sludges
- coal tar residues
- other miscellaneous residues

In 1976 the Hillman Company, via its foundation, donated the 35-acre downstream tip of Neville Island to Allegheny County for the purpose of developing a public recreation park (Figure 2). The Hillman Company is the parent corporation for the now-defunct Pittsburgh Coke & Chemical

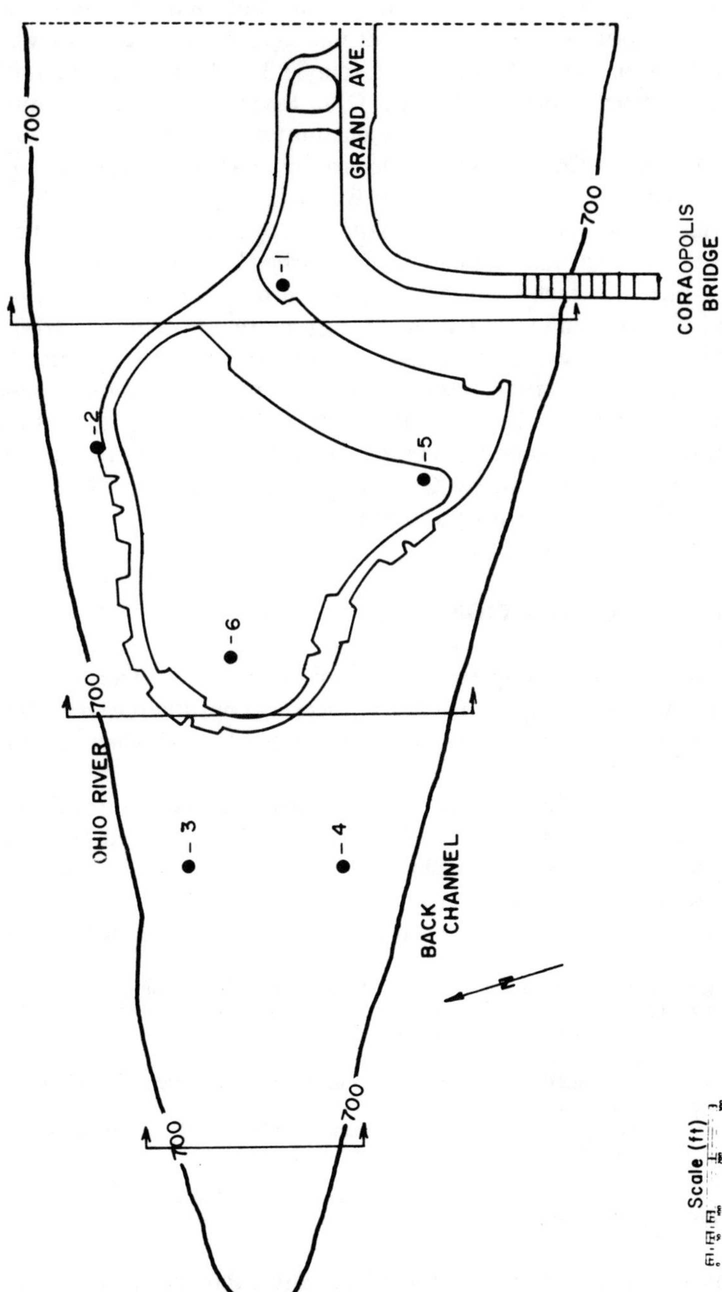

Figure 2. Location map of monitoring wells, sampling points and cross sections.

Company. The Allegheny County Department of Planning and Development was responsible for the overall management of the planning and construction for the site's conversion to a $3.3 million recreation facility. During the development of this site, Ohio River Park, various industrial residues and municipal wastes were encountered at and near the surface.

The Allegheny County Health Department (ACHD) became involved in this matter in July, 1978 when it was asked by the County Planning Department to investigate fumes emanating from certain areas of the park where buried waste material had been disturbed by construction activities. At this time, the park's construction was nearly complete and close to $2,000,000 had been expended. ACHD found that little information was available on the types and quantities of wastes deposited at the site. At the direction of ACHD, the County Planning Department commissioned a consulting firm to investigate these issues and to assess potential health implications. ACHD found this report inconclusive, and was subsequently charged by the Allegheny County Board of Commissioners with investigating the real or potential health hazard.

PHASE I INVESTIGATION

In response to the request by the Board of County Commissioners, ACHD conducted a search for an engineering consultant experienced in hazardous waste site investigations and selected Fred C. Hart Associates, Inc. (FCHA) of New York.

During this first phase, FCHA was charged with determining if a public health hazard existed at Ohio River Park. ACHD worked closely with FCHA on every phase of this Phase I investigation, which was performed during June and July of 1979.

The Phase I investigation specifically included the following:

1. gathering and interpreting historical background information;
2. mapping of surface wastes by visual inspection and delineating the likely locations of buried waste by application of electrical resistivity and metal detection survey techniques;
3. sampling and analysis of solid, liquid and gaseous wastes from the surface and near surface; and
4. assessing the toxicological aspects of the waste sampled and identifying possible exposure mechanisms.

Historical Background

As noted earlier in this report, the Ohio River Park site had been used as a waste disposal area prior to its development as a park. This was confirmed through discussions with former Pittsburgh Coke & Chemical employees and local residents. Other historical documents dating back to 1949 also supported this contention.

More specific information on some of the types, possible quantities and approximate locations of industrial waste were obtained from the preliminary review conducted for the County Planning Department's consultant and through interviews with the construction company personnel who built Ohio River Park. Discussions with these employees provided the following information:

1. Barrels were encountered throughout the major undercut areas. Several of the barrels were still filled with liquid wastes. These barrels were drained, crushed, and disposed in a burial pit.
2. Some of the waste material encountered in undercut areas were in a semi-solid state. This material was spread over the surface, and subsequently dried and solidified as the volatiles quickly evaporated.
3. Several construction workers left the job site. Apparently odors emanating from the site caused concern for their personal health.

Field Survey

A visual inspection of the park site revealed soil staining, vegetative stress, subsidence and a wide variety of waste exposed at the surface. The general topography of the site is basically flat, except for a distinctive mounding at the downstream tip. A closer inspection reveals this increase in elevation is most likely not a natural occurrence, but manmade.

The reported locations of subsurface wastes were further supplemented by the use of electrical resistivity and metal detection instrumentation. A number of areas of lower resistivity values (1500 ohm-ft) were identified, which may indicate the presence of waste or leachate from some waste. The metal detection scan proved unreliable because of the predominance of metal slag, which had been deposited at the park site. The results of this field testing, intensive visual inspection of the site and the historical background data were used to decide on exact sampling locations.

Sampling and Analysis

The sampling protocol included obtaining grab samples of buried wastes and liquids within 4 feet of the surface, plus any wastes exposed at the surface. Also, ambient air monitoring for potentially harmful vapors from the volatilization of buried and surface wastes was not feasible due to the interference of other sources of air pollution in the park's vicinity.

Therefore, it was decided to use the head space of surface and buried waste samples to test for the presence of any potentially harmful vapors.

The subsurface samples were performed by a backhoe digging test pits 4 feet deep at specified, predetermined locations (Table II). Samples from 15 of the 19 locations were sent for laboratory analysis. In a number of cases, decaying drums and leachate were encountered in some of the test pits.

Four major categories of chemical waste samples were collected in Ohio River Park:

1. pigments
2. coal tar residues
3. leachates, liquid organics and sludges
4. crystalline solids

Table III lists chemicals found in each type of sample. Discussions of the concentrations of these chemicals will be deferred until the following section.

Toxicological Assessment

The approach to this assessment was to identify the most hazardous chemicals detected in the samples in concentrations judged to be significant from a health standpoint. For those substances considered priority pollutants by EPA, water quality criteria and standards were used if available. Where such criteria were not applicable, other standards such as OSHA or Interim Primary Drinking Water Regulations were used. Table IV lists the toxic chemicals that are present at the park site in notable quantities.

A health risk assessment was done on the toxicological properties of six substances or classes of substances:

1. benzene
2. phenols
3. parathion
4. cyanide
5. mercury and
6. coal tar residues

This toxicological assessment was performed by the FCHA toxicologists and reviewed by technical staffs of EPA, Pennsylvania Department of Environmental Resources and ACHD. This assessment focused primarily on the identification of chemicals present. The chemicals' concentrations were considered baseline information to be used in assessing the health effects and not as an absolute indication of the present level. It is also

Table II. Location and Description of Sampling Sites

Location No.[a]	Location Description[b]	Sample No.	Depth[c]	Sample Description[d]
1	Near Picnic Deck D	1,1a	Surface	Blue wood chips and black tarry matter
		2	Surface	Orange granular clay-like substance
2	Bank by Ohio River	3	Surface	Black tar
3	Bank by Ohio River	4	Surface	Hard grey material
4	Bank by back channel	5	Surface	Tan solid, needle-like crystals
5	Bank of back channel	6	Surface	Mixture of white, rust and grey granular particles
6	Clump of trees	12	Surface	Hard, granular material (crystalline)
7	Disposal pit for excavated sludges	24	3.5	Brown soil
		25	3.5	Leachate
8	Bank by back channel	7	Surface	Blue wood chips
		8	3	Rock coated with blue crystals
		9	3	Soil
9	Spongy fill under proposed Nature Center	10	2	Black tar
10	Next to parking lot near proposed Nature Center	Not sampled		
11	Spongy fill under Picnic Deck P	11,11a	2	Black, viscous sludge
12	"Indian Mound"	15	1.5	Hard, blue crystalline material
		16	2	Black, tarry soil
		15	3-4	Black, tarry soil
13	Spongy fill under road	18	3	Tarry soil
		20	4	Leachate
14	Near Group Shelter E	21,21a	½-1	White and yellow crystals
		22	½-1	Blue-black, chalky crystals
15	Next to hydrant between Shelter I and Picnic Deck H	Not sampled		
16	Western tip of island	Not sampled		
17	South of oil well	13	2-4	Soil and tarry material
		14	Surface	Yellow resin
18	Between river stairs and parking lot	Not sampled		
19	West of loop road	23,23a	1	Black tar

[a]Sampling site location number—see map (Figure 5).
[b]Location description—physical features observed at sampling site.
[c]Sampling point depth—at or below ground surface.
[d]Sample description—visible appearance at time of sample collection.

Table III. Chemicals Found in Various Samples

I. COAL TAR RESIDUES	
A. Volatile Organics	Benzene Toluene Xylenes Trimethylbenzene Indane Naphthalene Methylindanes Dichlorobenzenes Trichlorobenzenes
B. Polycyclic Aromatic Hydrocarbons	Acenaphthalene Fluorene Phenanthrene Fluoranthene Pyrene Chrysene Anthracene
C. Other Organics	Dibenzofuran Phenols
II. PIGMENTS	
A. Volatile Organics	Benzene Toluene Xylenes Trimethylbenzene Naphthalene Dichlorobenzenes
B. Inorganics	Cyanide Copper Manganese Silicon Iron Magnesium Sulfur Aluminum Mercury
III. CRYSTALLINE SOLIDS	
A. Volatile Organics	Benzene Toluene Xylenes Trimethylbenzene Naphthalene Dichlorobenzenes Trichlorobenzenes
B. Other Organics	Phthalic Anhydride

Table III, continued

C. Inorganics	Silicon Iron Magnesium Aluminum Sulfur Cyanide
D. Pesticides	Parathion α-benzene hexachloride γ-benzene hexachloride (Lindane) Heptachlor
IV. LEACHATES	
A. Volatile Organics	Benzene Carbon tetrachloride Chlorobenzenes 1,1,2,2-Tetrachloroethane Chloroform 1,2-Dichloropropylene (*cis*) Ethylbenzene Tetrachloroethylene Toluene Trichloroethylene Xylenes Naphthalene Methylene Chloride Trimethylbenzene
B. Phenols	Phenol 2-Chlorophenol 2,4-Dichlorophenol 2,4,6-Trichlorophenol (*p*-Chlorometacresol) 4-Nitrophenol 2,4-Dinitrophenol
C. Polycyclic Aromatic Hydrocarbons (PAH)	Benzo(a)anthracene Benzo(a)pyrene 3,4-Benzofluoranthene Benzo(k)fluoranthene Chrysene Acenaphalene Anthracene Fluoranthene Phenanthrene Pyrene
D. Other Organics	Ethylbenzene N-Nitrosodimethylamine Pthalate esters N-Nitrosodiphenylamine

Table III, continued

E. Pesticides	Parathion α-Benzene Hexachloride γ-Benzene Hexachloride (Lindane) 2,4-D 2,4,5-T Silvex
F. Metals and Other Inorganics	Antimony Arsenic Copper Thallium Mercury Lead Nickel Manganese Cyanide

possible that other types of wastes with different chemical constituents were missed by the sampling program.

After assessing the health effects of the six substances, possible human exposure mechanisms were formulated. These mechanisms or scenarios were presented as "worst case" examples of hypothetical incidents should Ohio River Park be used as a recreation facility. Special attention was given to the special hazard presented by these wastes to children and women of childbearing age. Specific exposure scenarios were developed for:

1. benzene vapors;
2. explosion and fire hazard;
3. leachate exposure; and
4. skin irritants.

It was concluded from these scenarios that:

1. Benzene exposure potential is high.
2. The possibility of accumulation of flammable vapors poses a significant explosion and fire hazard.
3. Lethal and poisonous concentrations of hazardous chemicals could exist in the leachate.
4. Skin irritation or eye damage is possible from phenols and coal tar residues.

Table IV. Comparison of Toxic Chemical Levels Found at Site with Current Federal Criteria and Standards

Substance		Level			
		Air		Water	
Class	Compound	Highest Level Found at Site (ppm)	OSHA Standard (ppm)	Highest Level Found at Site (μg/l)	EPA Water Quality Criteria (μg/l)
Volatile Organics	Benzene	2066*	10	5,100	0**
	Toluene	85*	200	12,000	NA
	Xylenes	91*	100	—	
	Naphthalene	+	10	—	
Polycyclic Aromatic Hydrocarbons	Fluoranthene	—		150	200
	Benzo(a)pyrene	—		83	NA
Phenols	Phenol	—	5	3,700	1
	2-Chlorophenol	—	NA	800	0.3
	2.4-Dichlorophenol	—	NA	15,000	0.5
	2.4-Dinitrophenol	—	NA	136	NA
Chlorinated Solvents	Carbon tetrachloride	—	10	4	2.6
	Tetrachloroethylene	—	NA	5	2.0
	Chlorobenzene	—	75	8	NA
Pesticides	Parathion	—	NA	4,500	NA
Other Organics	Ethylbenzene	—	100	43	—
	N-nitrosodi-methylamine	—	0#	3	0.026
Metals and Other Inorganics	Antimony	—	0.5	400	NA
	Arsenic	—	0.5	100±	0.02
	Copper	—	NA	3,200	1.0
	Thallium	—	0.1	190	4
	Mercury	—	0.05	1,200	2 (proposed)
	Lead	—	0.2	230	50
	Nickel	—	1	160	NA
	Cyanide	—	40	1,320	5

Legend:
— Not detected in this medium.
* Head space analysis results.
+ Present in head space, but only qualitative data available.
NA Criteria or standard not available.
Carcinogen (only recommended limit).
± Reliable concentration not available.
** For maximum protection of human health.

Conclusion Phase I Investigation

FCHA concluded from the Phase I investigation that a public health hazard did exist at Ohio River Park. It stated in its July 1, 1979 report that:

> Considering the chemicals present at the park site, their observed prevalence, measured concentrations, toxicological and synergistic properties and the possible exposure mechanisms of park visitors to those chemicals, users of the park could experience adverse health effects.

PHASE II INVESTIGATION

As a result of the Phase I investigation, FCHA was commissioned to undertake a second study in the fall of 1979. This followup study was undertaken to determine potential offsite environmental effects resulting from the release of park contaminants; to investigate areas of Ohio River Park that appeared to be free of contamination; and to develop possible remedial alternatives. This effort was accomplished through:

1. the installation of eight ground water monitoring wells;
2. analysis of surface and ground water samples;
3. conducting electrical resistivity and metal detection surveys on eastern portion of site; and
4. additional interviews with individuals in light of Phase I study results.

Installation of Monitoring Wells

Eight ground water monitoring wells were installed in conjunction with Ohio River Park. Six of the wells were for the Allegheny County and two were requested by EPA.

The six county wells were located at the park site. Four of the county wells were located at the site's perimeter to check for lateral migration of contaminants offsite. The other two were installed on the park's interior to check for any ground water contamination directly under the site and any ground water mounding. One EPA well was placed on the easternmost border of the park, while the other was located approximately 1500 feet east of the park (Figure 2). These wells were dug to check for offsite migration of contaminants both to and from the park site.

Each of these wells was installed to provide:

1. a means of sampling ground water;
2. a means of measuring ground water levels;

3. a means of obtaining soil samples;
4. information on subsurface materials at the site; and
6. sampling points for a ground water monitoring program.

Figure 3 displays the type of monitoring well that was used at Ohio River Park.

During the drilling of the test borings, split spoon soil samples were obtained at 5-foot intervals for visual classification and laboratory analysis. From the sieve tests conducted, grain-size distribution curves were used to estimate the permeability of the sand and gravel aquifer. In general, the subsurface conditions encountered by the test borings are shown in Figure 4 and Table V. Typically, the subsurface conditions revealed a sand and gravel aquifer 40 feet thick and, in most areas, covered with alluvial deposits 20 feet thick, consisting of silty fine sands and clays.

It should be noted that three test borings (MW-2, 3 and 4) encountered buried wastes; and of these borings, two (MW-2 and 4) also encountered voids. Samples of this waste were collected and submitted for analysis.

Priority pollutant analysis was performed on each of the county's four perimeter wells' ground water samples (MW-2, 3, 4 and 5). MW-1 and MW-6 were only analyzed for volatile organics. The EPA's "offsite" wells were analyzed for only representative constituents of the major wastes known or suspected to exist at the park site.

Field Survey of Eastern Sector

Both electrical resistivity testing and metal detection scans were conducted at the lesser investigated eastern portion of the site. The resistivity testing results suggest the existence of less "fill" material than the rest of the park (Figure 5). Furthermore, other than discovering a large deposit of slag, no other significant readings were recorded during the metal detection scan (Figure 6). These tests support an initial contention that the 10-acre eastern portion of the site is most likely uncontaminated. This contention was based on the fact that a chain link fence, which had separated this eastern portion of the park from the downstream section for many years, may have acted to restrict the area of waste disposal.

Additional Interviews

In light of the data accumulated during the Phase I investigation, followup interviews were conducted with personnel who supervised the construction of the park. Figure 7 delineates the generic types and

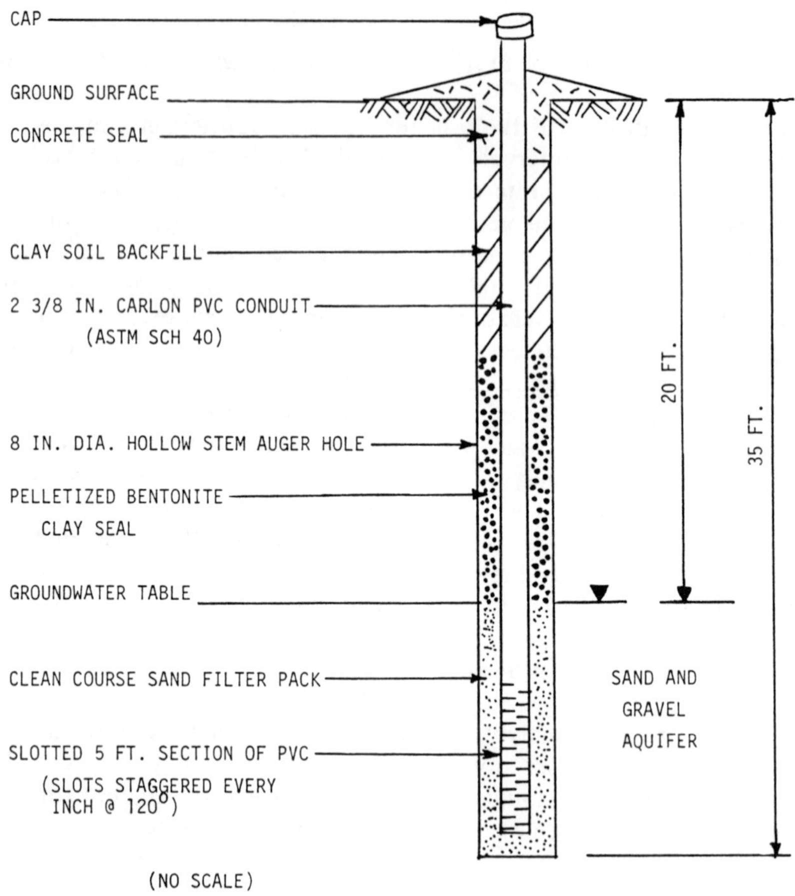

Figure 3. Typical monitoring well construction.

locations of materials encountered during the park's excavation and construction. It is important to note that a large amount of the park's surface was disturbed during site construction. Consequently, wastes in several areas were displaced, contributing to a greater irregularity in disposal patterns, a probable patchwork distribution of toxic wastes, and further contamination of virgin soils.

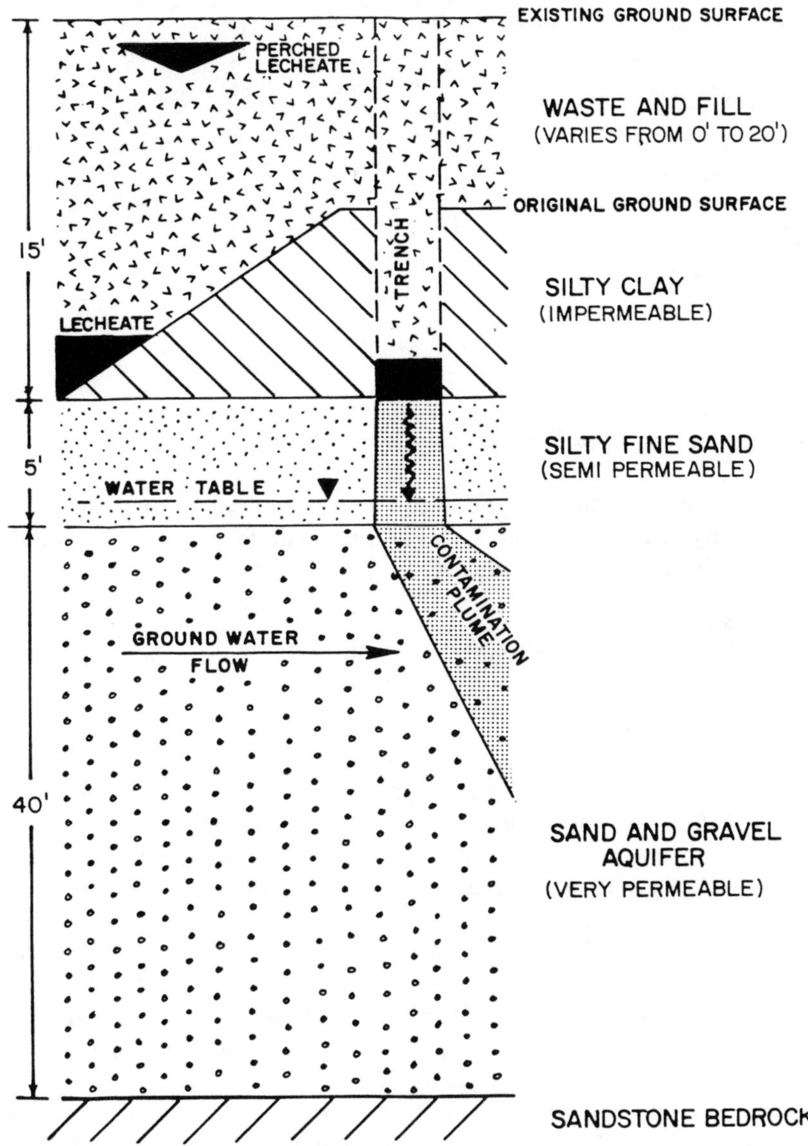

Figure 4. Schematic of typical subsurface conditions (no scale).

Table V. Monitoring Well Locations and Water Surface Elevations

MW	Station Location	Existing Grain-Size Elevation	Natural Grain-Size Elevation	Water Surface Elevations 9-29-79	11-8-79	12-6-79	12-7-79
1	87+35,14′R	716.6	716.6	695.7	695.5	695.9	695.7
2	23+20,38′L	715.2	705.2	696.0	695.7	695.7	695.6
3	31+62,32′L	723.3	708.3	696.0	?	695.3	695.1
4	44+70,25′L	721.4	710.9	695.9	695.9	696.2	696.1
5	52+11,47′L	719.8	719.8	696.0	695.5	695.9	695.6
6	80+72,77′L	716.9	716.9	696.0	695.4	695.6	695.6
1-A	91+12,35′R	715.4	715.4	695.6	695.4	695.7	695.6
2-A	Offsite, 1500 East	?	?	?	?	?	?

Surface and Ground Water Analysis

The results of the ground water analysis reveal that a number of contaminants exist at the park that exceed Clean Water Act Water Quality Criteria. These contaminants are:

- metals
- volatile hydrocarbons and chlorinated hydrocarbons
- phenols
- cyanide

The most serious ground water contamination occurs at MW-4, both with respect to the number of contaminants and their concentrations (Figures 8–11). Table VI outlines the ground water contaminants that exceed the water quality criteria for human health. The contaminants that exceed the human health criteria include a number of substances considered to be potential human carcinogens (arsenic, beryllium, benzene, chloroform, trichlorethylene, PCB, *bis*-(2-chloroethyl) ether and 1,3-dichloropropylene).

The samples analyzed from EPA's wells showed little evidence of contamination from the park site. Only higher than normal arsenic, lead and mercury levels were detected.

Samples were also analyzed of the ponded water on site and storm sewer discharges from the park to the Ohio River. Both the storm sewer discharges and the ponded water contained small quantities of parathion and its degradation by-products. The detection of these substances in the runoff indicates that the source of these wastes is somewhere in the catchment area and near the surface.

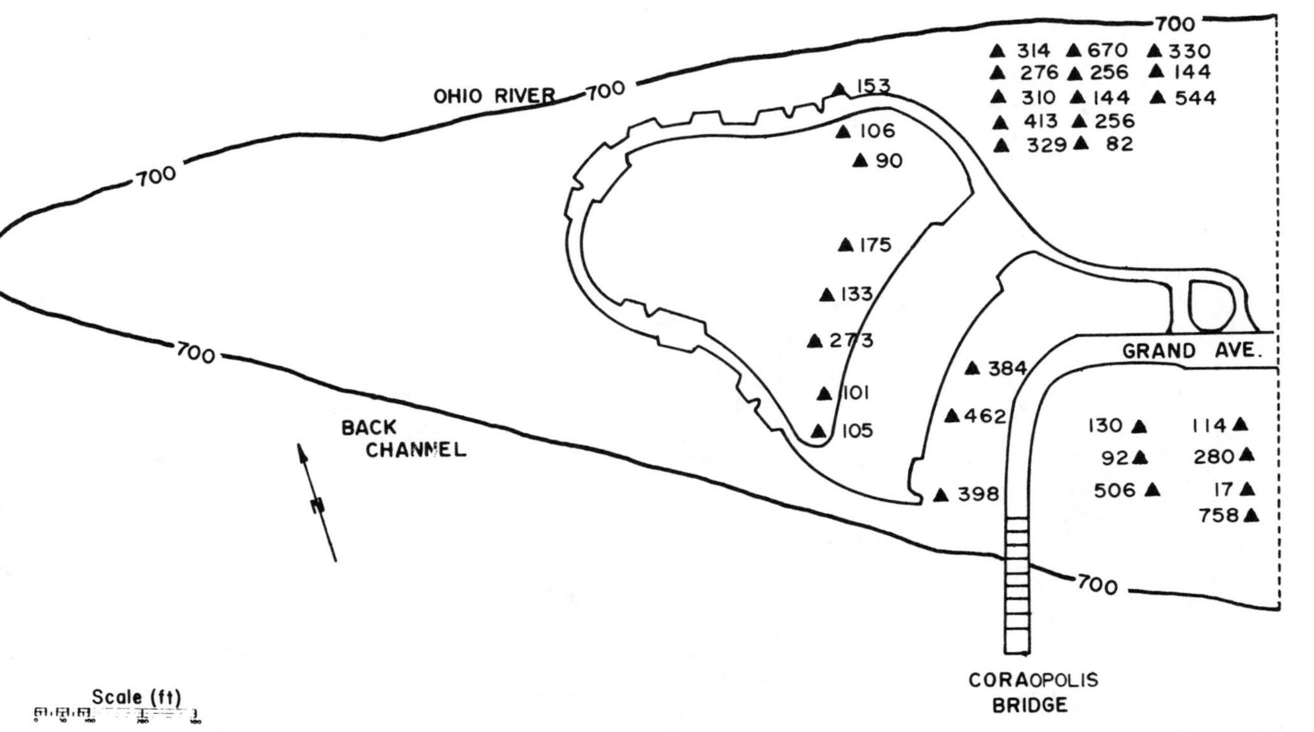

Figure 5. Soil resistivity map.

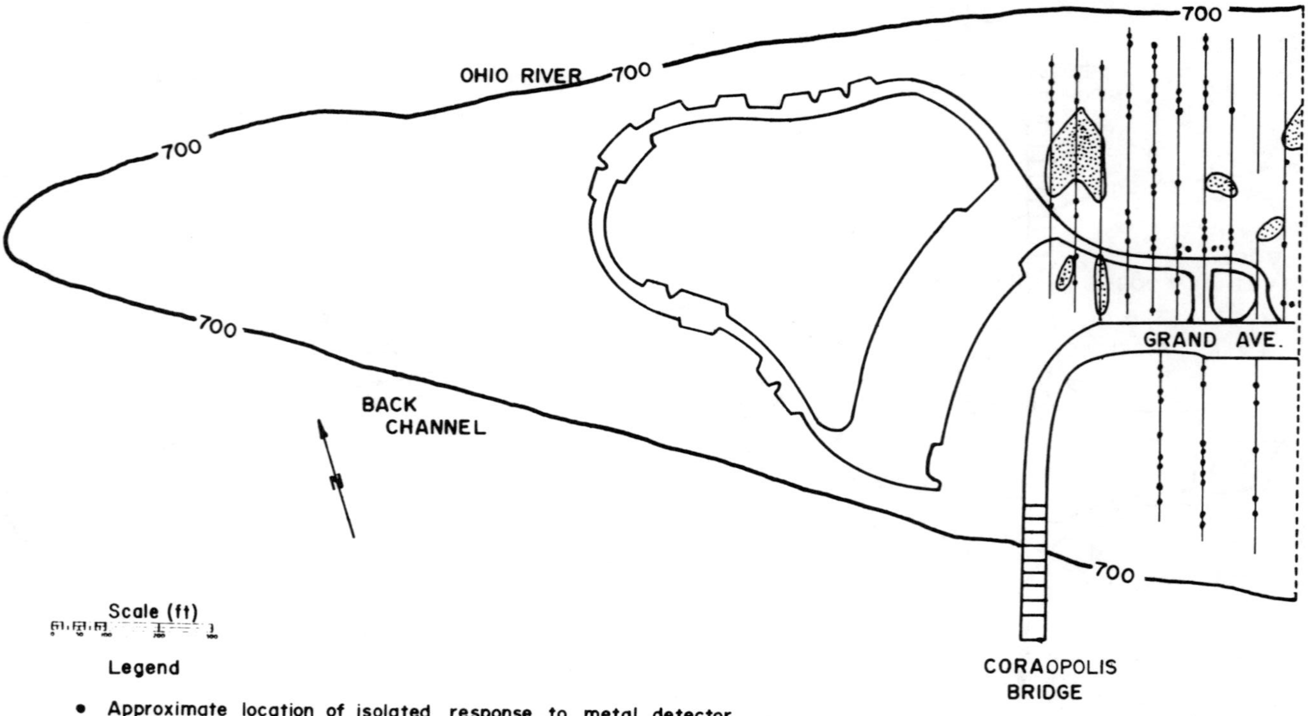

Figure 6. Metal detection map.

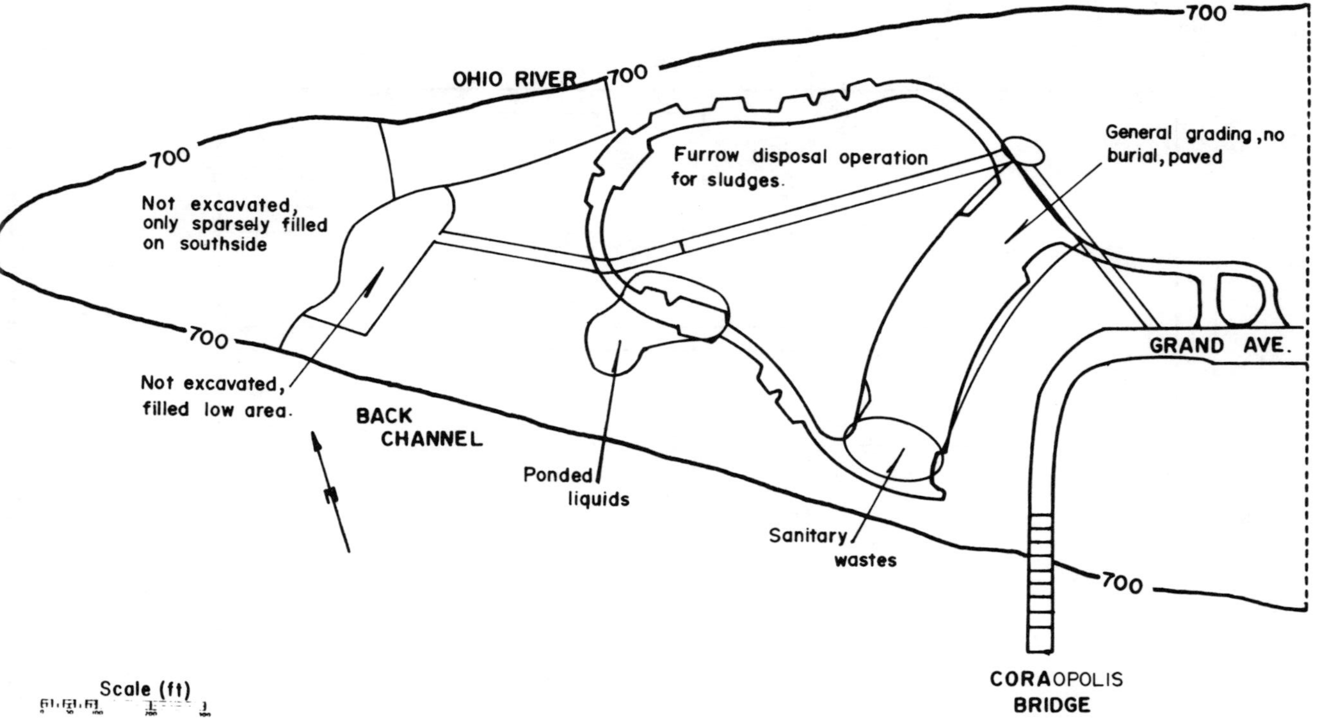

Figure 7. Generic descriptions and locations of materials encountered during construction.

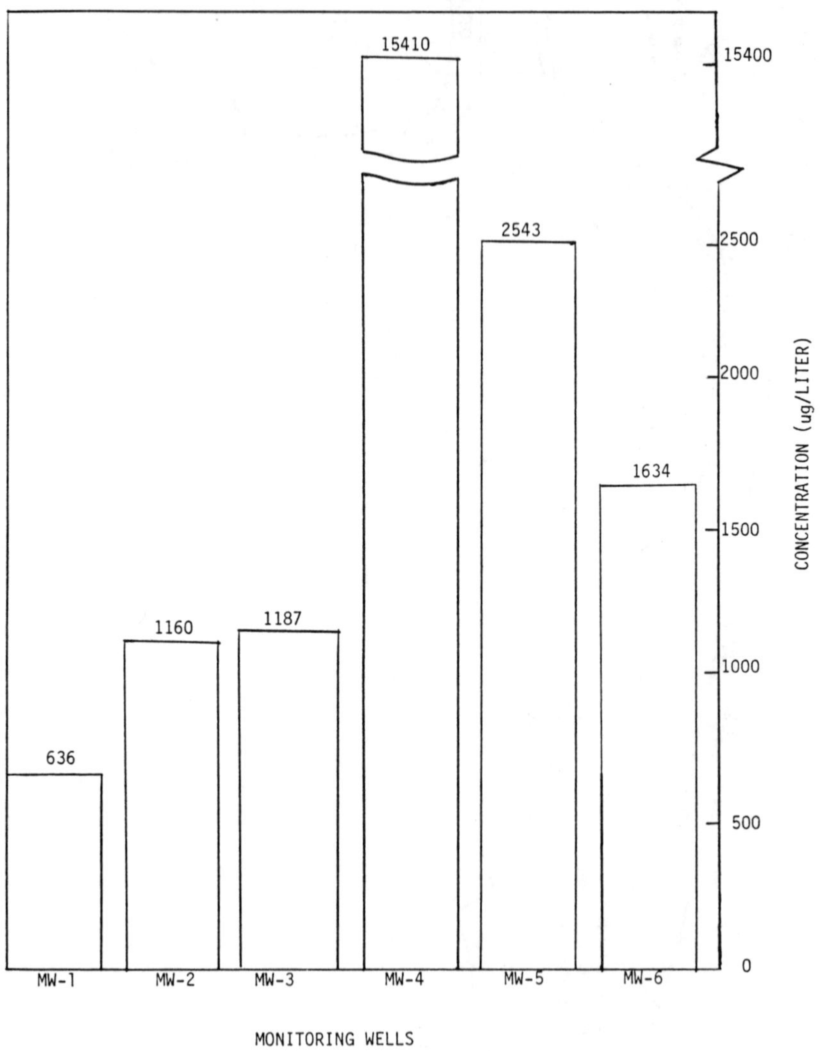

Figure 8. Total concentration of metals in ground water samples.

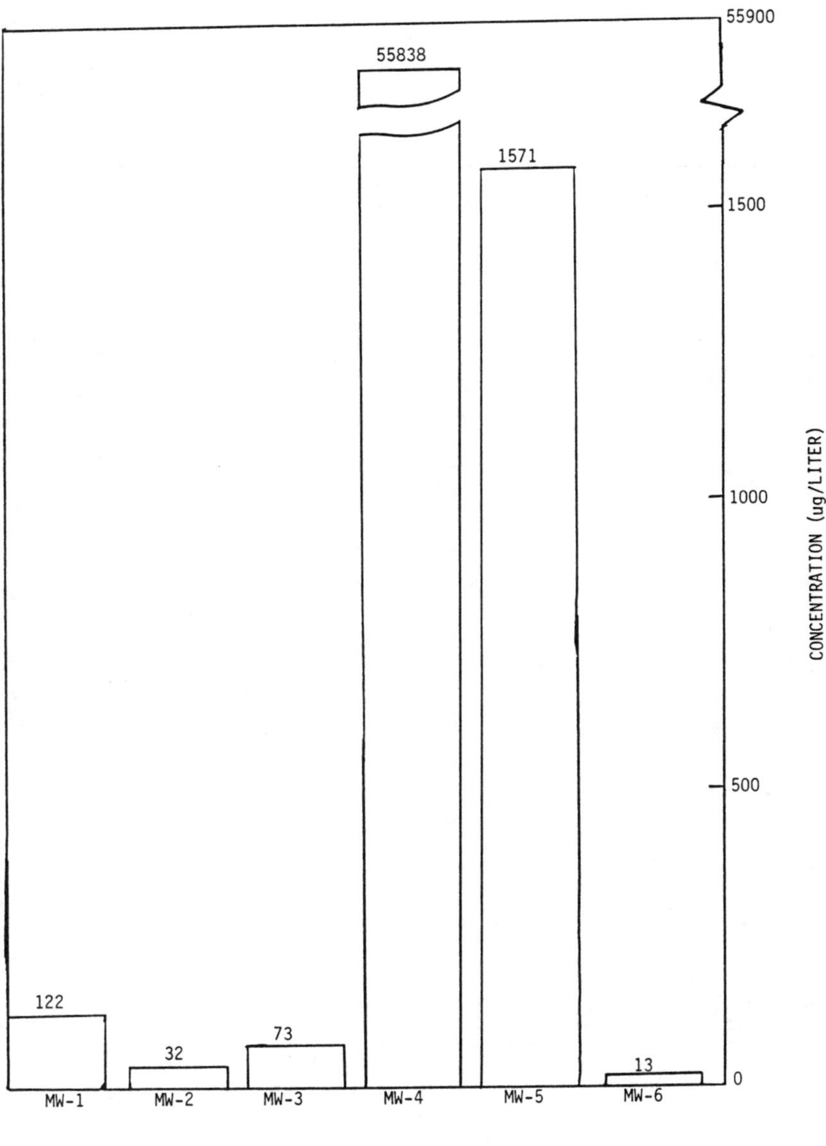

Figure 9. Total concentration of volatile organics in ground water samples.

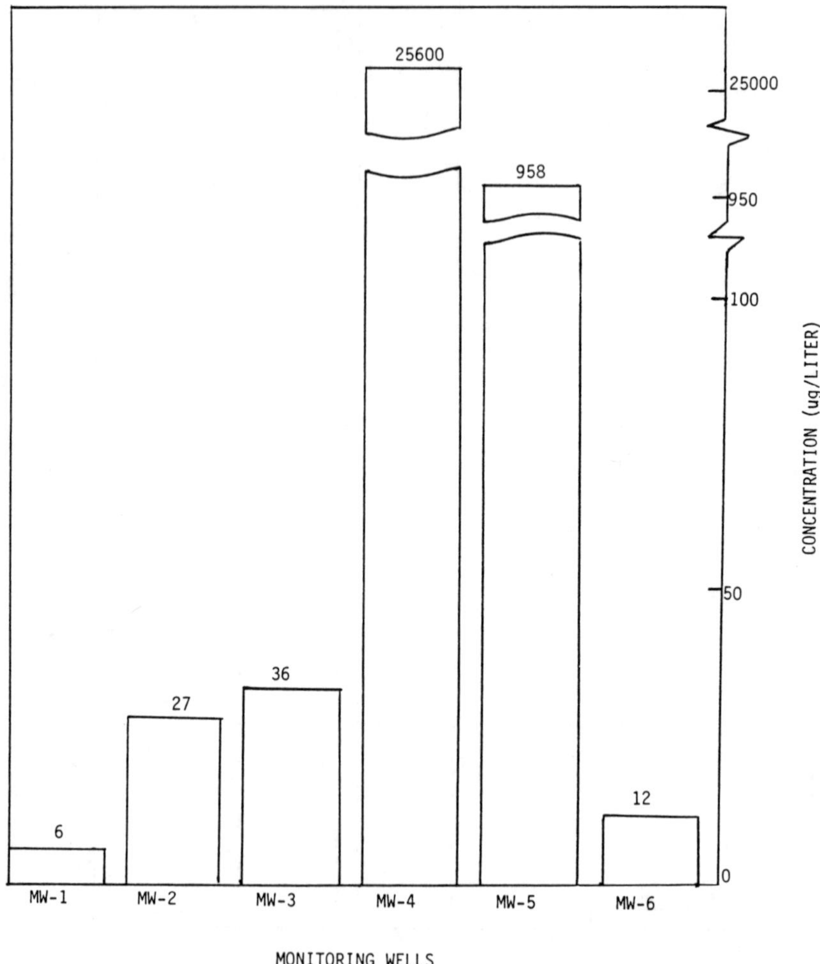

Figure 10. Concentration of phenolic compounds in ground water samples.

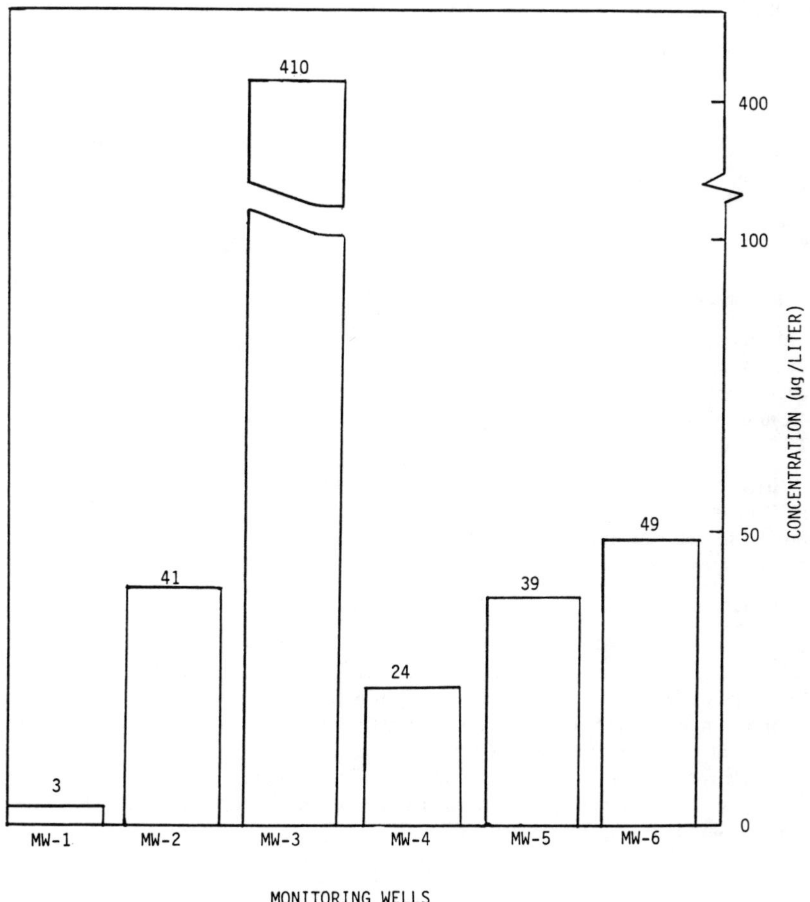

Figure 11. Concentration of total cyanide in ground water samples.

Table VI. Ground Water Contaminants Exceeding Water Quality Criteria for Human Health[a]

	Ratio of Concentration in Sample to Criteria Concentration					
Contaminant	**MW-1**	**MW-2**	**MW-3**	**MW-4**	**MW-5**	**MW-6**
Cyanide			2[a]			
Arsenic	300[b]		825[b]		1650[b]	700[b]
Cadmium				7		
Chromium Total				3		
Copper				1.3		
Lead	1.6		2.8	4.2	8.2	3.6
Nickel	1.1	5.1		22	2	1.5
Silver				6		
Antimony		1.4	2	6	2	2
Beryllium				1379[b]	230[b]	
Thallium				75	25	25
Zinc				1.9		
Benzene			2.5[b]	3200[b]	102[b]	
Naphthalene				3.6		
Chlorobenzene				3.6[c]		
Chloroform				81[c]		
Trichloroethylene	2.8[b]			1.8[b]		
2-Chlorophenol				137	30	
4-Chlorophenol					3	
2,4-Dichlorophenol		12	10	440	58	
PCB-1254			1000[b]		3200[b]	
PCB-1248			950[b]		2700[b]	
bis (2-Chloroethyl) ether					52[b]	
1,3-Dichloropropylene						4.8
Methylene Chloride	8.5	3.5	4.5			
4.4′-DDT		428	20			
2,4-D				22[d]		

[a]Criteria are for free cyanide. Sample analysis is for total cyanide.
[b]For carcinogens, the ratio of concentration in the sample to the concentration corresponding to the interim target risk level of 1 per 10^5 is given.
[c]Criteria for tainting (20 μg/l) used for calculation of ratio.
[d]Ratio of concentration in ground water to maximum contaminant level (MCL) specified by the Safe Drinking Water Act.

Waste Distribution

Estimates of waste volume and distribution were necessary to assess various mitigative measures and associated cleanup costs. This was partially accomplished by comparing the change in contour elevations on the 1941, 1964 and 1978 topographical maps of Neville Island. Nineteen cross-sectional figures were drawn perpendicular to the main axis of the island (Figure 2). Three comparative examples of this technique are shown in Figures 12–14.

To determine the actual quantities of waste deposited at the site, the cross-sectional areas of fill were calculated. The 1941 contours were assumed to be virgin soil and the baseline for this procedure. The volumes were computed using the "average end area" method. A total of approximately 225,000 cubic yards of fill was computed using this procedure. If this material were distributed evenly over the 35 acres of the site, the average fill depth would be over 5.5 feet.

These comparisons and calculations also were used to develop a contour map displaying the approximate depth to natural soil (Figure 15). Also superimposed on this map are the general locations and areal extent of different waste types determined through sampling and analysis, data from the construction log of the site and other field data.

In most instances, the wastes were *not* encountered in discrete locations, as indicated by Figure 15. Instead, they were in a heterogeneous, unsegregated mass, which was further mixed by the extensive redistribution of soil and waste during the park's construction. Because of this fact, it may be impossible to isolate the hazardous wastes from the nonhazardous fill. An approximate areal and volumetric distribution of the various waste types is presented in Table VII.

Figure 16 provides the waste classifications based on the sampling points, soil test pits and borings performed at the park site. Three material classifications were selected:

1. Natural soils and "suspected" clean fill or "probably uncontaminated materials," e.g.,
   (a) Municipal solid waste
   (b) Slag
   (c) Foundry sand
   (d) Random coal, bricks, gravel and wood
2. "Probably contaminated material," e.g.,
   (a) Miscellaneous industrial wastes
   (b) Tar
   (c) Chemically soaked wood and material;
3. "Confirmed contaminated material," e.g.,
   (a) Pesticides
   (b) Coal tar residues
   (c) Phenols
   (d) Volatile organics

Figure 17 provides a qualitative evaluation of the "degree of hazard" for the entire park site. The assessment is based on the aggregation and predominance of waste types in a given location. Almost 10 acres of the

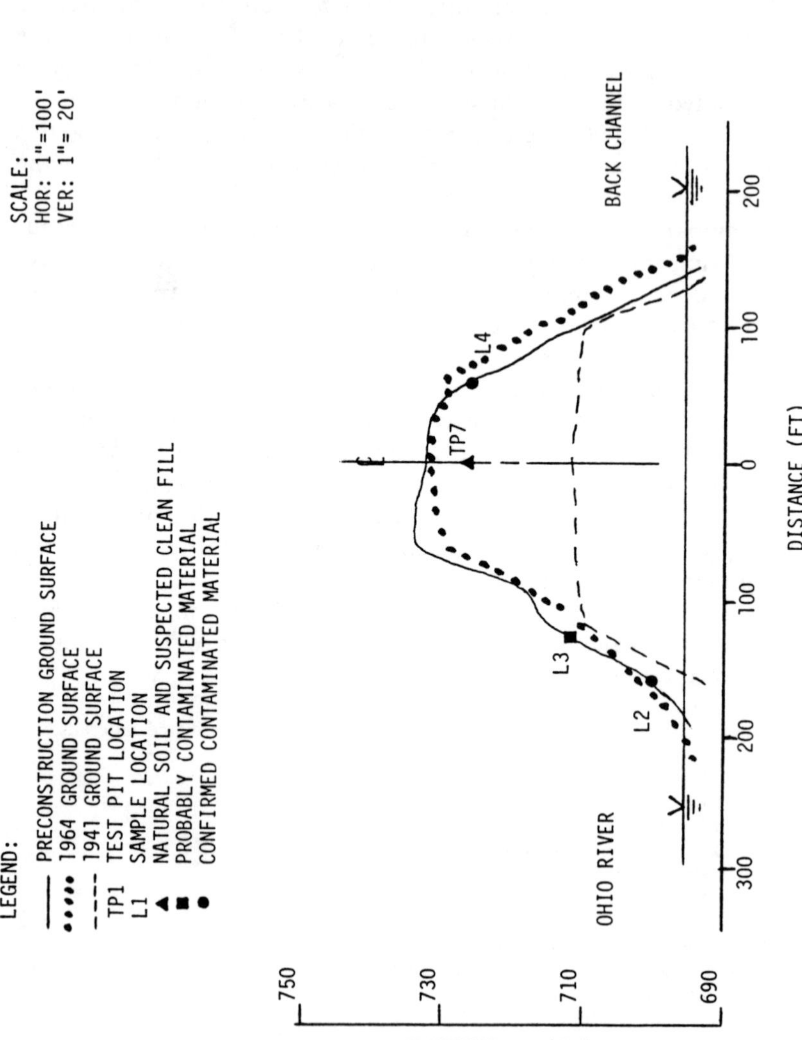

Figure 12. Cross section station 71+00.

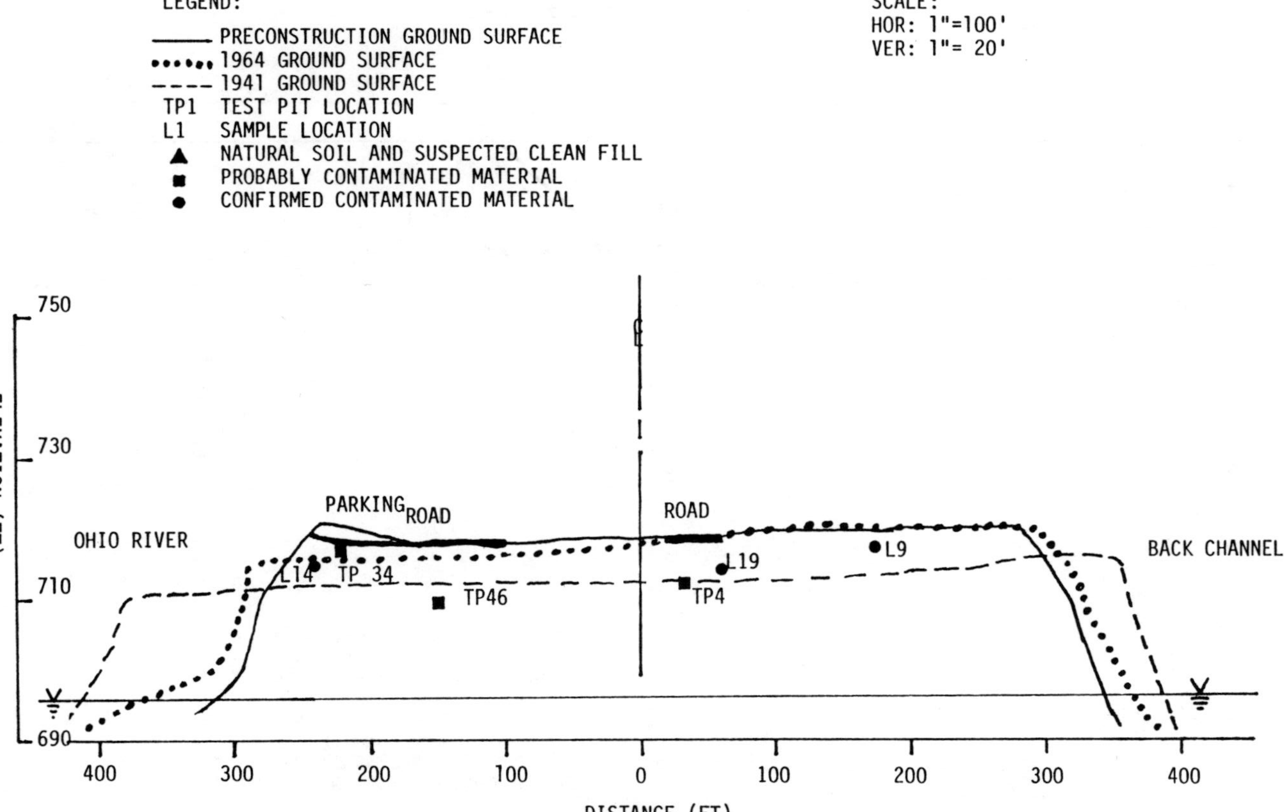

Figure 13. Cross section station 79+45.

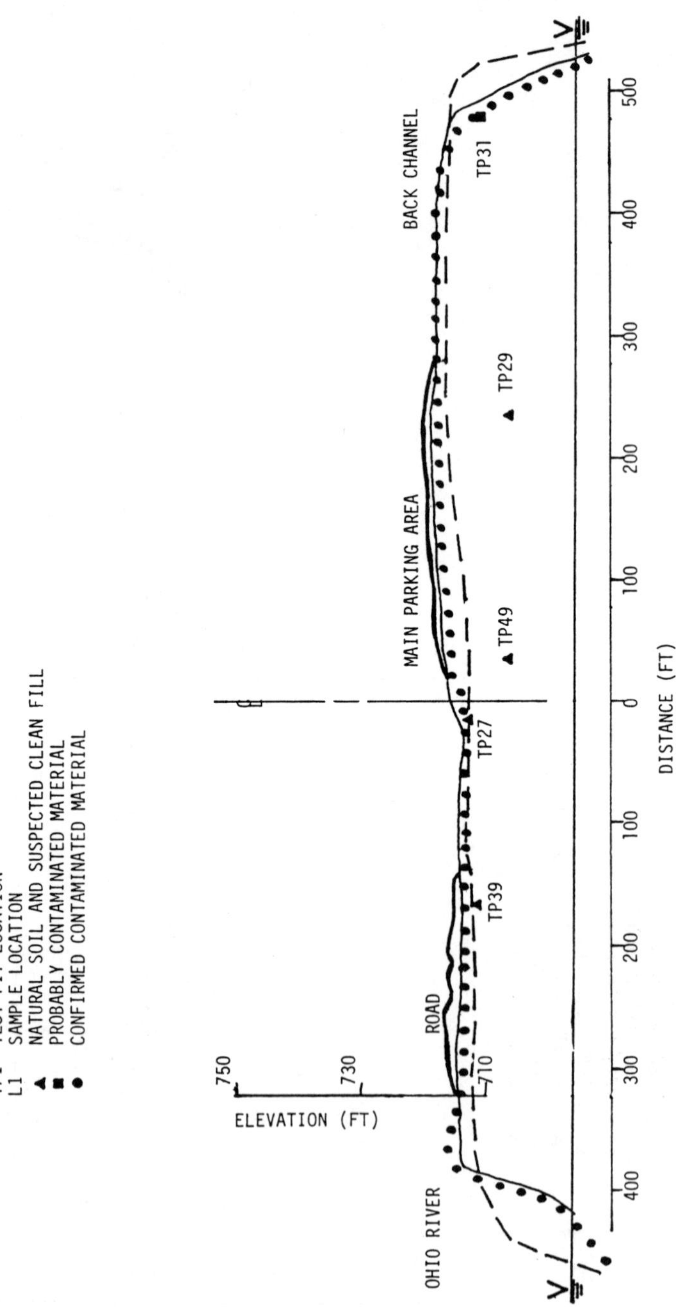

Figure 14. Cross section station 86+00.

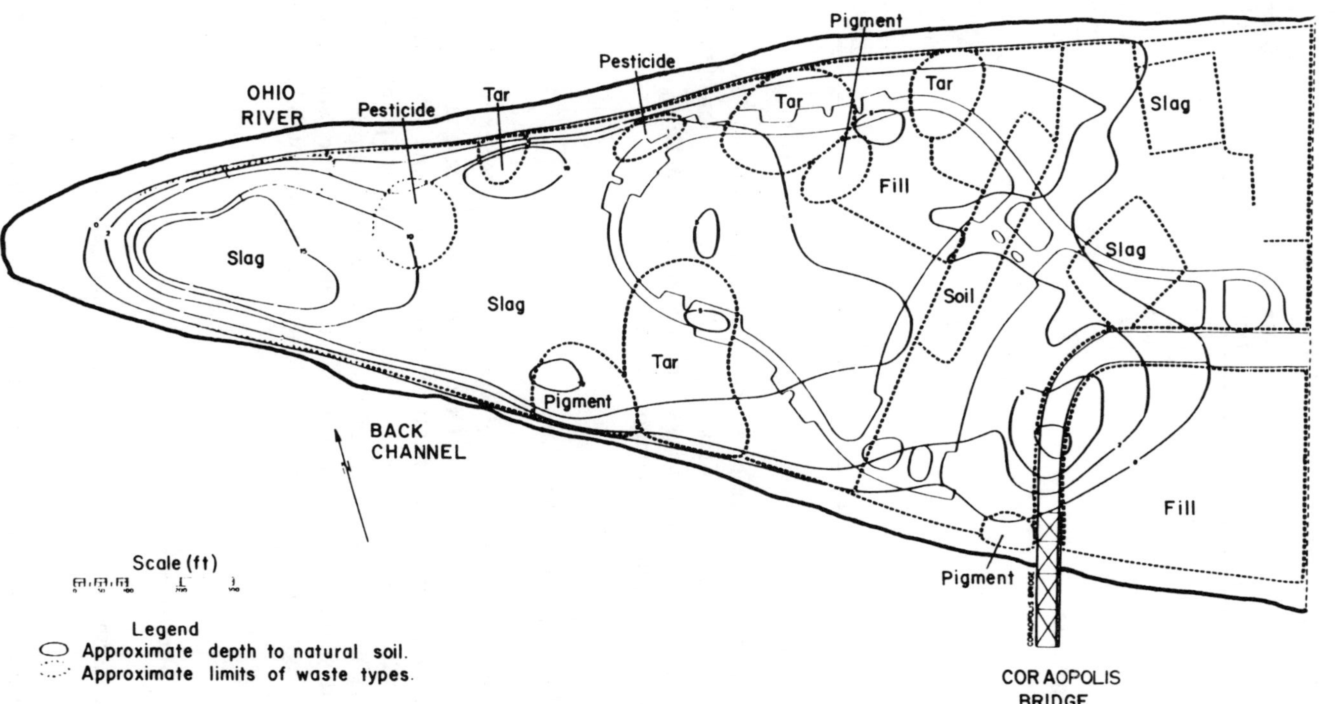

Figure 15. Waste disposal areas and depth of fill.

Table VII. Approximate Distribution of Waste Types Disposed on Property

Type of Waste	Area (ac)	Quantity (yd^3)
Hazardous		
Industrial waste	1.3	16,800
Pesticides	0.25	4,700
Tar	3.4	32,600
Pigments	1.2	10,900
SUBTOTAL	6.1	65,000
Nonhazardous		
Slag	11.6	90,400
Miscellaneous fill	10.8	50,400
Soil	6.2	19,200
SUBTOTAL	28.6	160,000
TOTAL	34.7	225,000

park contained wastes classified as "confirmed contaminated," and more than 7.5 acres are "probably contaminated." The eastern section of the site and the areas along the river banks, which comprise 17 acres of the park, are denoted as "probably uncontaminated."

Phase II Investigation Conclusions

It was concluded that the surface water runoff, storm sewer outfalls and ground water discharge from Ohio River Park release a slow but continuous flow of contaminants downstream into the Ohio River. However, when taken together with all of the upstream point sources discharging into the Ohio River and its tributaries, it is not possible under the limited scope of this study, to accurately determine whether this increased degradation perceptively interferes with the potability of the river.

Mitigative Alternatives

After assessing all available data, three mitigative alternatives were developed:

1. Continue park closure and abandon plans for the park at a cost of \$147,000–\$248,000.

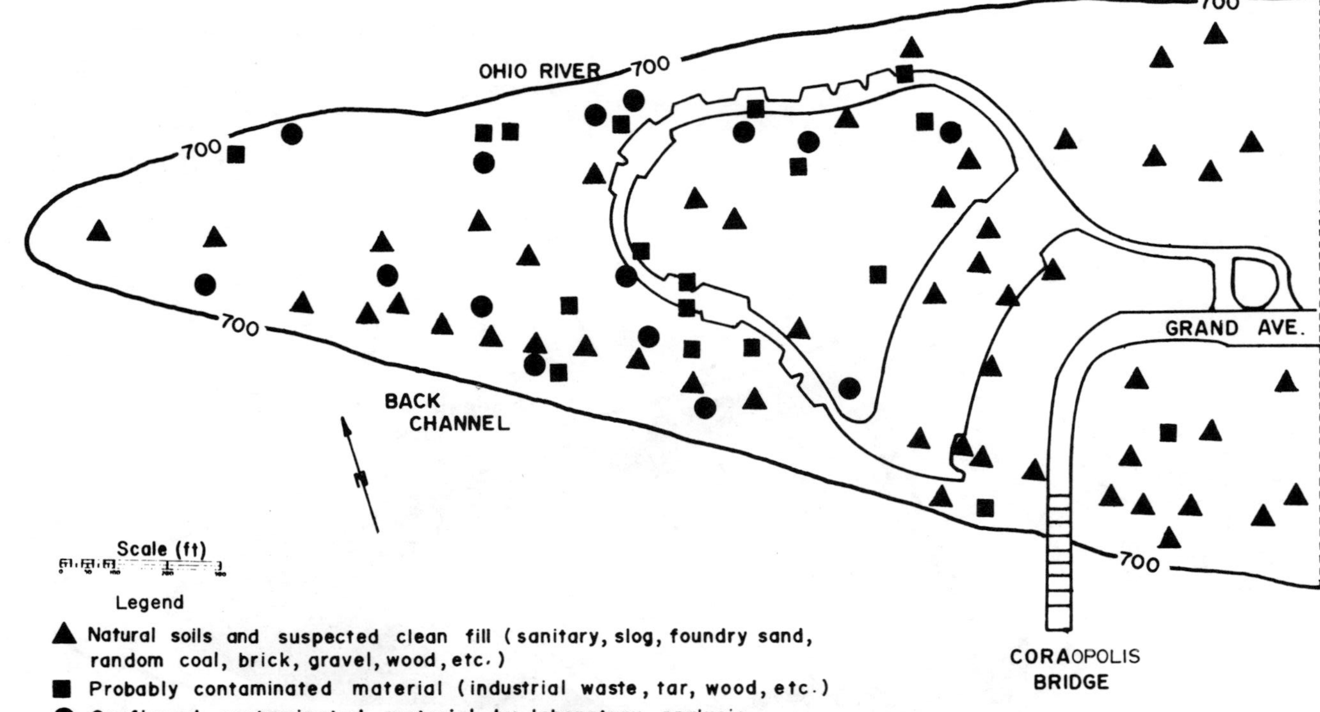

Figure 16. Waste classification based on sampling points, soil pits and borings.

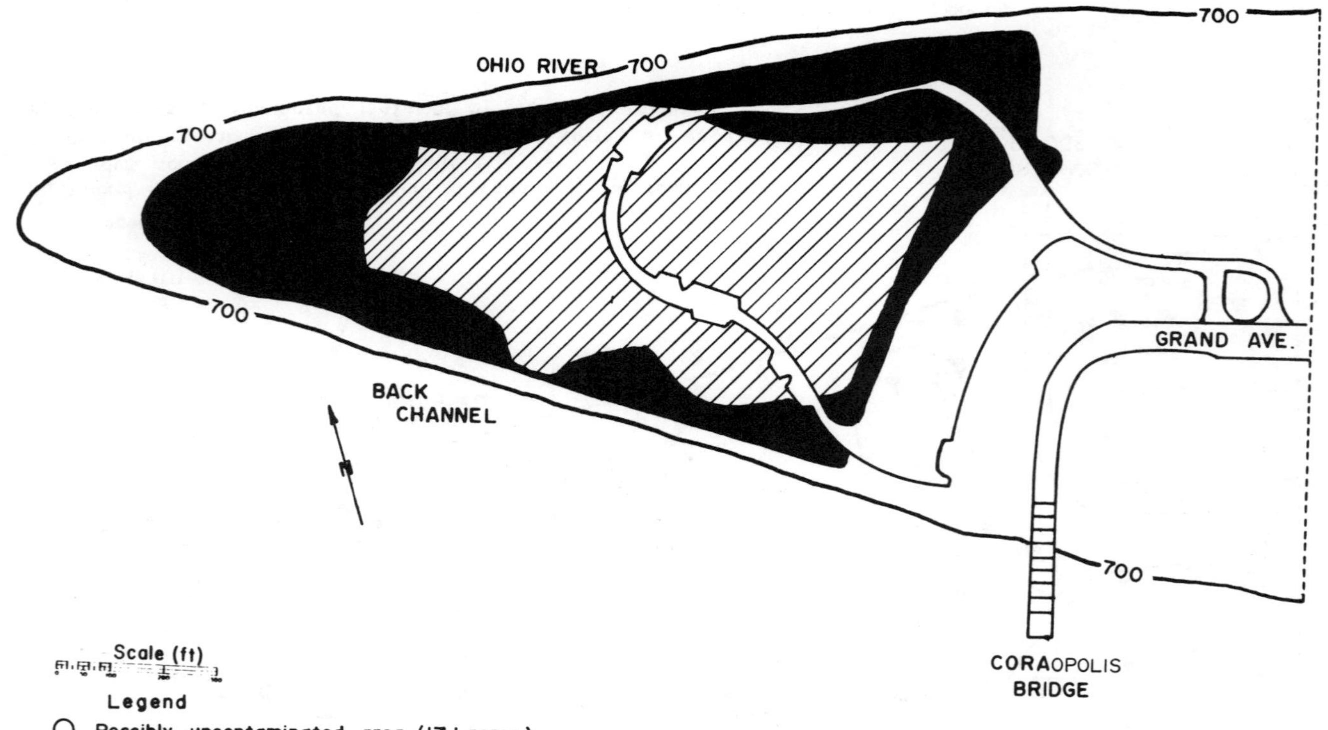

Figure 17. Area assessment map.

2. Remove all contaminated wastes and soil to open entire park at a cost of \$ 6,900,000–\$ 24,000,000.
3. Develop a limited portion of the park on the possibly uncontaminated eastern portion of the park at a cost of \$ 296,000–\$ 430,000.

Of course, a fourth option was available to the county that was not included in FCHA's Phase II report. The county could try to negotiate the return of the park site to the Hillman Foundation, the original donor of the land. In January of 1980 the Allegheny County Board of Commissioners, on reviewing the results of FCHA's second study, decided to pursue the fourth option.

In the spring of 1980, the Hillman Foundation agreed to take ownership of the Ohio River Park property back from the county. Moreover, it also agreed to reimburse the county for the funds expended in both developing the park and subsequently investigating its hazards. Hillman has stated that it plans to hire another private engineering consulting firm to both monitor and reassess the situation at the park.

SUMMARY AND FUTURE CONSIDERATIONS

Although the site ownership has been returned to the Hillman Foundation, ACHD continues to be concerned for the safety of the public drinking water supplies in the vicinity. In its present state, the park remains a significant potential menace to the public health of downstream water users. Figure 18 and Table VIII display and outline the location and general operational data for these water supply treatment facilities. Additional comprehensive sampling, monitoring and assessment of these facilities and the Ohio River will be necessary to assess the park's role in contributing to degradation of the river's water quality. For such an effort, the technical assistance and resources of the Pennsylvania Department of Environmental Resources and the EPA will be necessary to properly address this issue.

Regardless of what course of action is taken by Hillman on this matter, a monitoring and sampling program of the water supplies should be implemented. Furthermore, any onsite remedial actions taken should be performed with worker safety in mind. Disturbance of a waste burial area is always an experience full of many unknown hazards and surprises. Also, precautions must be taken to protect the residents in the area from the release of harmful vapors or hazardous leachate at the site if subsurface work is undertaken.

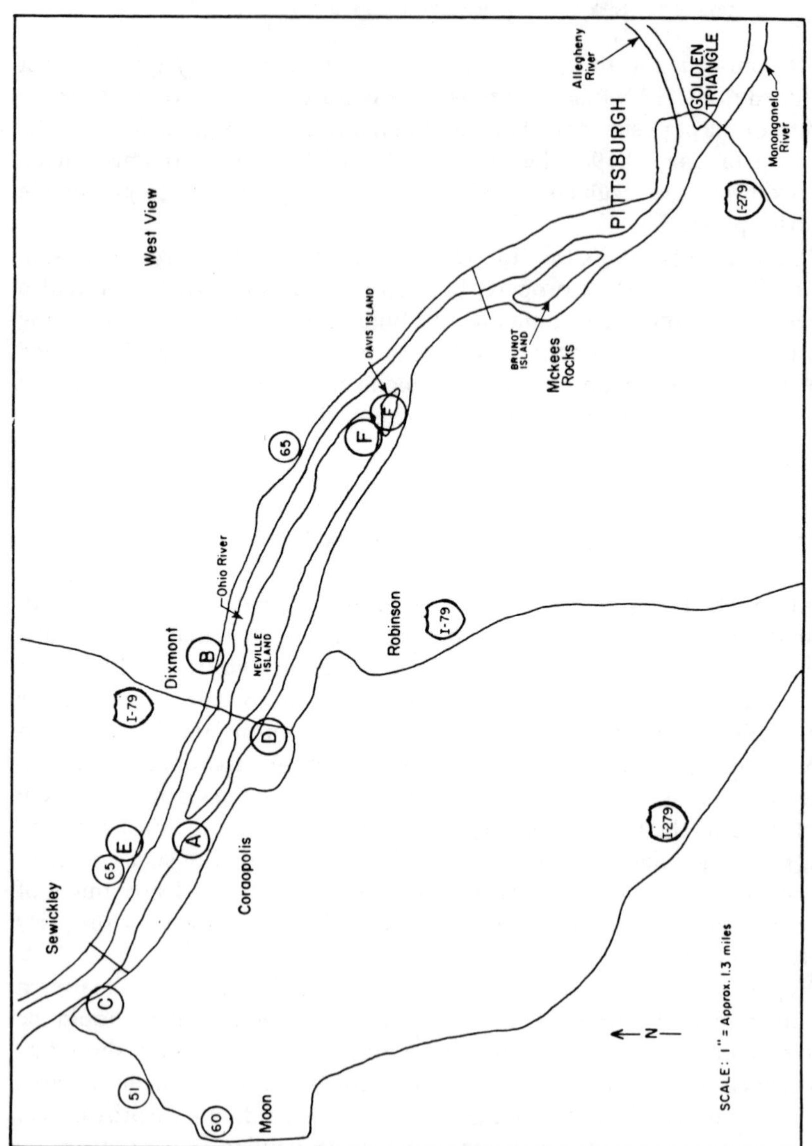

Figure 18. Municipal water users in the Neville Island vicinity.

Table VIII. Municipal Water Users Near Neville Island

Map Symbol	Water System (municipality)	Average Flowrate (mgd)	Intake	Treatment
A	Coraopolis	1.2	Wells	Filtration Chlorination
B	Dixmont	0.2	River	Chlorination Filtration Powdered activated carbon
C	Moon	2.5	Wells	Chlorination Filtration Fluoridation
D	Robinson	1.5	River	Filtration Chlorination Fluoridation Powdered activated carbon, if necessary
E	Sewickley	1.0	River, Wells	Filtration Chlorination Fluoridation
F	West View	15.5	Wells	Filtration Chlorination Fluoridation

From this overall investigation of Ohio River Park, a number of general conclusions can be reached:

1. Tests and test procedures do not now exist to determine the effects of chemical waste on man.
2. Comprehensive toxicological data regarding chemicals in waste disposal sites are not now available.
3. Adequate information and procedures for testing complex mixtures of waste residues and chemical by-products are not now present.
4. There is a great need for more environmental toxicologists in the public and environmental health professions.
5. The greatest risk of an abandoned dump site is probably not from the direct exposure to the waste chemicals, but through contamination of drinking water.
6. The types, quantities and concentrations of different wastes existing at disposal sites are difficult to determine. Furthermore, toxicological information does not exist for most of them.
7. Many of the waste residues and chemicals are not commercial products but their precursors or process intermediates. Consequently, very little chemical or toxicological research has been done on these manufacturing by-products.

CHAPTER 11

EXPERIENCE OF THE ACCIDENT AT SEVESO, ITALY*

L. Bisanti, F. Bonetti, F. Caramaschi, G. Del Corno, C. Favaretti, S. Giambelluca, E. Marni, E. Montesarchio, V. Puccinelli, G. Remotti, C. Volpato and E. Zambrelli

Regione Lombardia
Seveso, Italy

On July 10, 1976, at noon, a runaway reaction started in a plant for the synthesis of trichlorophenol (TCP) at the ICMESA chemical factory in Meda, a city 15 miles north of Milano. The safety valve of the reactor was not provided with protective devices, so the hot vapors blew into the atmosphere and the resulting cloud, moved by a weak wind, began to fall to the ground covering an area southeast of the plant. A few days later, lesions of the burning type began to appear on the skin of several people, mainly youngsters; and grazing animals, mainly rabbits, began to die. It was only nine days later that ICMESA admitted that tetrachlorodibenzodioxin (TCDD), and not only TCP, sodium hydroxide, sodium trichlorophenate and ethyleneglycole, could have been expelled from the reactor, so that for about two weeks the people were not submitted to any protective measure against the damages caused by TCDD.

---

*Reprinted from the Proceedings of the 6th ETS Conference, published in 1979 by Akadèmiai Kiadò, Budapest by permission.

Immediately, the County Health Laboratory went to work trying to define the extent and intensity of the TCDD pollution by means of gas chromotography-mass spectrometry (GC-MS) determinations in superficial soil samples taken at 50- to 100-meter intervals.

Three zones were then established, as seen in Figure 1. Zone A was subdivided in 7 subzones—A1–A7, with A1 being the closest to the factory and the most polluted. Zone B is located southeast of zone A, while zone R ("Respect"), which surrounds both zones A and B, was originally judged free from contamination. Subsequently, as the sensitivity of the test improved, it was found to be slightly and unevenly contaminated.

Table I shows the size of the area involved, the average and top concentrations of TCDD for all zones and subzones. Average TCDD concentrations were highest in subzone A1 (580.4 ug/m^2) and lowest in zones B (3 ug/m^2) and R (0.9 ug/m^2); but an uneven distribution of TCDD, particularly in zones B and R, is suggested both by the high ranges of individual measurements and by the percentage of points, for each zone, found "free" (threshold concentration: 0.75 ug/m^2) from contamination (from 24.5% in zone B up to 68.6% in zone R).

Immediately after preliminary data were obtained, people were evacuated from zone A; on the other hand, inhabitants of zones B and R were left there and asked to follow a number of hygienic rules, including temporary advice neither to start pregnancies nor breast feed babies. Only evacuation of children and pregnant women was provided. Growing vegetables and breeding animals were strictly prohibited; orchards were destroyed; and all edible animals were taken away or killed after their owners were compensated.

Therefore, inhabitants of zone A are to be considered as a group exposed directly to the toxic cloud and then to the most polluted area, but only for as long as 19 days. Could the inhabitants of zone B, who were left in their homes, be considered as having been chronically exposed to a much lower amount of TCDD?

Whether such chronic exposure is real or only theoretical depends on the following:

1. How does TCDD in the soil reach its organisms? Should all precautions suggested by the health authorities be followed? This is a point of great debate, which has yet to be clarified.
2. How carefully do they follow the precautions suggested? This is very hard to state. For instance, due to a lack of grossly detectable health consequences, many people never stopped or have resumed illegally growing vegetables and breeding animals. By the end of August 1978, 100 people in Zone B were charged in court with disobeying the mayors' ordinances not to resume breeding animals and growing vegetables [1].

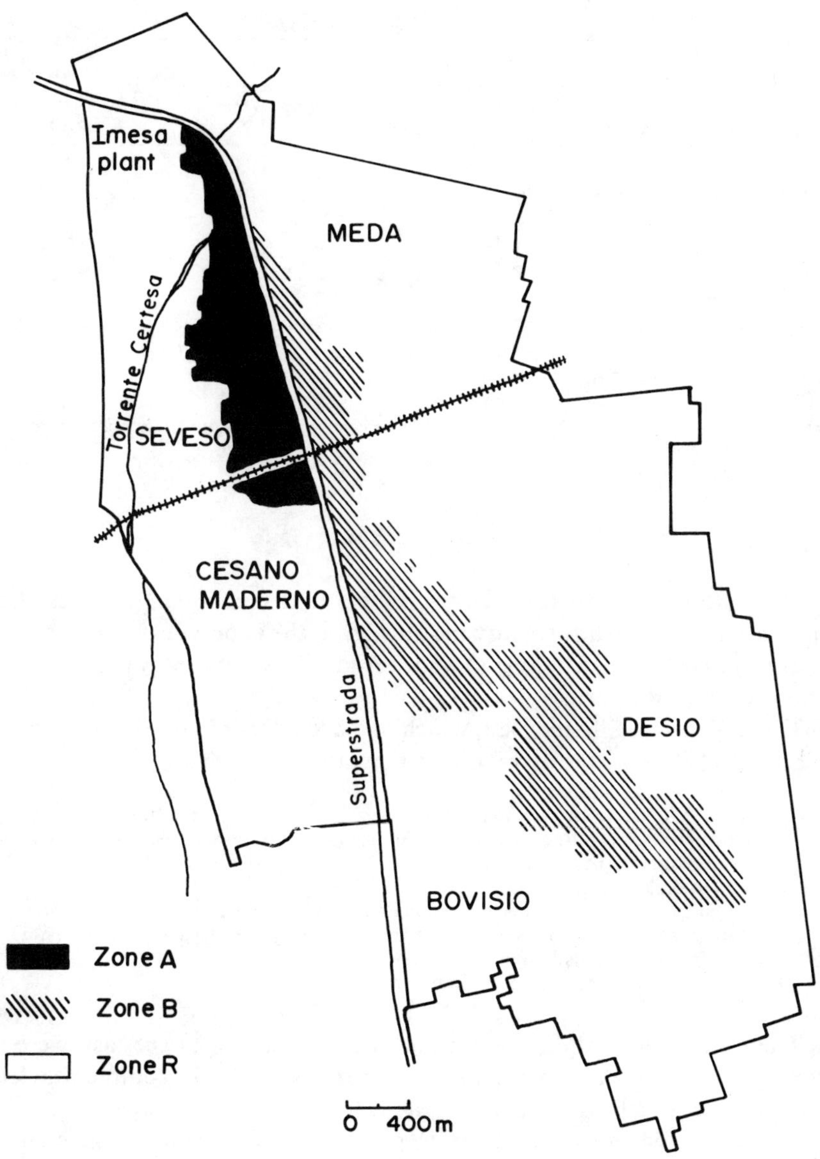

Figure 1. Schematic map of Seveso area by pollution zones.

Table I. Distribution of TCDD Contamination ($\mu g/m^2$) in the A, B and R areas on the basis of GC-MS analysis of soil samples (data provided by Regione Lombardia, Piano Operativo no. 1)

Zone	Size (ha)	TCDD Values ($\mu g/m^2$) Average	Top	n° Samples	Negative Samples[a] No.	%	Estimated Amount of TCCD on the Area (g)
A tot	80.3			306	12	*3.9*	147.5
A 1	10.7	580.4	544.7	51	1	1.9	62.1
A 2	5.1	421.1	1700	19	0	0	26.5
A 3	9.2	350.5	2015	34	3	8.8	32.2
A 4	7.2	134.9	902	26	3	11.5	9.7
A 5	16.3	62.8	427	50	2	4.0	10.02
A 6	14.0	29.9	270	61	2	3.2	4.1
A 7	17.8	15.5	91.7	65	1	1.5	2.7
B	269.4	3.0	43.8	106	26	*24.5*	8.0
R	1430.0[b]	0.9	9.7	449	308	*68.6*	8.5

[a]Less than 0.75 $\mu g/m^2$.
[b]Only 950 ha mapped.

In addition, due to too many political and ideological debates that circulated in the area for many months, even the exposure questionnaires administered to those people of zones A and B who agreed to cooperate, often were hard to trust.

The health surveillance plan, which had been illustrated two years ago in Gargnano [2] and slightly modified thereafter, included:

1. clinical followup of each individual living in the polluted area, not mainly for epidemiological purposes, but because of humanitarian reasons, and as it had been requested by the population;
2. longitudinal control of groups at higher risk
3. special projects, searching for damage of a neurological, immunological and chromosonal nature as suggested by previous experience of accidents in humans and by experiments in animals; and
4. surveillance of general health indicators.

The objectives of the program and the importance of the task were in contrast with the reality of the situation, namely an almost complete lack of local public health organization.

In spite of that, the decision of Regional Council was not to give direct responsibility in the plan to outer organizations, but to improve the local offices and complete their staff. This decision was democratically correct and in agreement with the health philosophy of the region; however, it took a tremendous amount of time to have the plan started and well underway.

So, after the generous help given by many on a volunteer basis during the very initial months, things progressed too slowly. For instance, the epidemiological working team only began in January 1978! Therefore, data collection and elaboration of available data are still far from being complete.

Data to be presented includes information about birth defects, abortions, births and deaths in the area. We shall also present data on the only clinically ascertained consequence of the accident, chloracne, and give a summary of the results of some special projects underway.

Table II shows the distribution by zone of the population involved in the accident. Zone A includes segments of the population of two cities, Seveso and Meda; zone B includes the segments of three: Seveso, Cesano and Desio; and zone R, 31,800 people, the segments of six (the four quoted, plus Seregno and Barlassina).

According to Table II, the 11 cities can be divided into three groups. The first group, Seveso and Cesano, had more than 50% of its population included in the contaminated area (A + B + R); the second group (Meda and Desio) had an involvement of around 20%; the third group (7 cities) was only 0.2% involved.

CHLORACNE

During the first month after the accident, only lesions due to the burning components of the cloud were observed among the inhabitants of the area; afterward up to 187 cases of chloracne were diagnosed: 164 in children 0–14 years old and the other 23 in adults.

A first wave of chloracne, 50 cases (34 of them children 0–14 years), was detected from September to December, 1976; 137 more cases (only 7 among adults) were diagnosed between February and April, 1977, including both people referred by physicians and those found during an ad hoc school screening.

Distribution of cases and rates by pollution zones is shown in Table III. Rates for zone A of both (1) "early" and (2) "late" chloracne were by far the highest. But for "late" chloracne, the rates for Seveso and Meda, zone R, were higher than for the whole of zone B.

The presence of some chloracne cases in areas where soil pollution is poor, scattered or apparently absent, is hard to explain, also because answers to exposure questionnaires are often incomplete or hard to verify. Remembering that Zone R of Seveso and Meda are closer to zone A than many parts of southern Zone B, people from such zones could have had direct or indirect contact with Zone A before it was evacuated and fenced off.

Table II. Distribution of Inhabitants (No. and %) by City and Pollution Zone[a]

Cities	No. (and %) Inhabitants					
	Total	Zone A	Zone B	Zone R	Zone A+B+R	Zone non A,B,R
Seveso	16,975	668 (3.93)	628 (3.69)	7,945 (46.79)	9,241 (54.45)	7,734 (45.55)
Cesano M.	33,799	—	2,736 (8.09)	14,991 (44.35)	17,727 (52.44)	16,072 (47.56)
TOTAL 1	50,774	668 (1.31)	3,364 (6.62)	22,936 (45.17)	26,968 (53.11)	23,806 (46.89)
Meda	19,571	62 (0.31)	—	4,017 (20.52)	4,079 (20.83)	15,492 (79.17)
Desio	33,011	—	1,373 (4.15)	4,608 (13.95)	5,981 (18.10)	27,030 (81.90)
TOTAL 2	52,582	62 (0.11)	1,373 (2.61)	8,625 (16.40)	10,060 (19.14)	42,522 (80.86)
Bovisio M.	11,225	—	—	167 (1.48)	167 (1.48)	11,058 (98.52)
Barlassina	5,656	—	—	72 (1.28)	72 (1.28)	5,584 (98.72)
Seregno	36,838	—	—	—	—	36,838 (100)
Lentate	13,037	—	—	—	—	13,037 (100)
Varedo	11,841	—	—	—	—	11,841 (100)
Nova M.	19,467	—	—	—	—	19,467 (100)
Muggio	18,690	—	—	—	—	18,690 (100)
TOTAL 3	116,754	—	—	239 (0.20)	239 (0.20)	116,515 (99.80)
TOTAL 1+2+3	220,110	730 (0.33)	4,737 (2.15)	31,800 (14.47)	37,267 (17.84)	182,843 (82.06)

[a]The six cities whose territory was partially classified in one or more of the pollution zones (A,B,R) are members, with other cities, of three Health Administrative Departments (Consorzi Sanitari di Zona or C.S.Z. in italian). Therefore, under zone "non (A,B,R)", in the last column, five cities are listed that belong to the three C.S.Z. but that are 100% outside the pollution zones (e.g., Seregno, Lentate, etc.).

Table III. Chloracne Cases (No. and 1000 inhabitants) by Pollution Zones

		(1) Chloracne Sept-Dec 1976 "Early"		(2) Chloracne Feb-Apr 1977 "Late"		Chloracne (1) + (2)	
		No.	**x1,000**	**No.**	**x1,000**	**No.**	**x1,000**
Zone A		46	63.01	15	20.55	61	83.56
Zone B		0	—	9	1.90	9	1.90
Seveso	Zone R	1	0.13	28	3.52	29	3.65
Meda	Zone R	0	—	20	4.98	20	4.98
Cesano M.	Zone R	0	—	13	0.87	13	0.87
Desio	Zone R	0	—	2	0.43	2	0.43
Seveso	Zone non(A,B,R)	0	—	13	1.68	13	1.68
Meda	Zone non(A,B,R)	0	—	14	0.90	14	0.90
Cesano M.	Zone non(A,B,R)	0	—	8	0.50	8	0.50
Desio	Zone non(A,B,R)	0	—	5	0.18	5	0.18
Other Cities		3		10		13	
TOTAL		50		137		187	

Table IV. No. and Percentage of Subjects 0–14 Years Showing Dermatological Lesions [zones A, B, R, non(A,B,R)]

		Zone A		Zone B		Zone R		Zone non(A,B,R)	
Dermatological Lesions		**No.**	**%**	**No.**	**%**	**No.**	**%**	**No.**	**%**
1. Chloracne, Sept-Dec 1976 ("Early")	to	31	14.5	0	0	0	0	0	0
2. Chloracne, Feb-Apr 1977 ("Late")	to	11	5.1	8	0.5	63	0.7	46	0.1
3. Chloracne (1) + (2)		42	19.6	8	0.5	63	0.7	46	0.1
4. Other lesions[a]		2	0.9	7	0.5	38	0.4	44	0.1
5. TOTAL of subjects 0-14 yr[b]		214	—	1,468		8,680		48,263	

[a]Atrophodermy Feb.-Apr. 1977 + dermatological lesions before July 10, 1976 + atrophodermy before July 10, 1976.
[b]Population 0–14 years on December 31, 1976.

Table IV shows the distribution by pollution zone of cases (and rates) of "early" and "late" chloracne and also of "other lesions" which include atrophodermy and dermatological lesions (referred as dating back before July 10, 1976) in children 0–14 years. "Early" chloracne apparently hit only children from Zone A*; the highest rate of "late" chloracne was in zone A and the lowest in Zone non(A,B,R), while in Zone B and R the "late" chloracne rates were low and almost the same. Speculation reported for the previous table can be repeated. In addition, we must consider the "other lesions," whose distribution in Zones A, B and R is not different from that of chloracne. The significance of those lesions and their distribution by zone, taking into account that production of TCP by ICMESA, dates back to 1975 and is now under careful consideration.

Table V shows the relative risks of chloracne in children by pairs of zones. Relative risks of zone A vs B, R and non R are highest; relative risks of B and R vs non(A,B,R) are still around 7, relative risks of B vs R is, on the other hand, close to 1. Other studies are in progress on people who contracted chloracne.

From clinical and dermatological reports it appears that those people had no other signs of pathology until now. Also the immune system of a group of children with chloracne showed no signs of alteration within two years after the accident, compared to the controls [3].

BIRTH DEFECTS

Just after the accident of July 10, 1976, fearing that TCDD could be responsible for a rise in the rate of birth defects (BD) (as suggested by the animal literature), special recommendations were issued to local physicians, stressing the importance of complying with notification of BD,

Table V. Chloracne (1) + (2) in Children 0–14 Years, Relative Risk for Pairs of Groups (Zones), Standard Error and Confidence Limits at 95% Probability Level

		Confidence Limits 95%	
Relative Risks (RR)	**S.E. (ln RR)**	**Lower**	**Upper**
A vs B = 44.6	0.39	20.6	96.5
A vs R = 33.4	0.21	22.0	50.8
A vs non(A,B,R) = 255.9	0.23	164.1	339.1
R vs B = 1.33	0.38	0.6	2.8
B vs non(A,B,R) = 5.7	0.38	2.7	12.2
R vs non(A,B,R) = 7.7	0.19	5.2	11.2

*Only 31 out of 34 cases of "early" chloracne reported in children 0–14 years are considered here. Three more children, who live in other cities (two in Milano, one in Mariano Comense) became involved while spending their vacations in Zone A.

which is mandatory. This was done because it was well known that BD in Italy are largely under-reported: Table VI shows that only 1 or 2 BD out of 1600–1800 births were reported, per year, between 1972 and 1975, in the four cities that were to become the most polluted by TCDD. The corresponding rates (0.55–1.24 × 1000) were not too far away from the national rates, 1.6–1.9 × 1000. It has also been ascertained that before the ICMESA accident many physicians had agreed not to report several malformations because such notification was felt by the parents as a socially negative mark. (A retrospective study is now underway to retrieve BD born in previous years by means of child screening and examination by cause of death certificates.)

During 1976 only four BD were reported, three before July and one in August, obviously all of them unrelated to TCDD: the expectation of the possible BD rise was for the beginning of January 1977, when women up to three months pregnant at the time of the accident were to deliver their babies.

During 1977 (with no clustering in any particular month) 38 BD were reported in the 11 cities, 7 in Seveso and Cesano, 16 in Meda and Desio, 15 in the other seven cities. During the first semester of 1978, an additional 22 cases of BD were reported. Rates for the three groups of cities are, respectively, 12.72, 23.02 and 9.81 in 1977; 16.7, 21.8 and 13.97 in 19.78 (Table VI).

Table VI. Birth Defects in the Areas Involved by the ICMESA Accident and in Italy, 1972–1978 (I–VI)

Year	Seveso + Cesano		Meda +Desio	Other 7 Cities	Italy
1972	1/1712	or	0.58 °/oo	NA[a]	1.8
1973	1/1812	or	0.55 °/oo	NA	1.6
1974	1/1721	or	0.58 °/oo	NA	NA
1975	2/1603	or	1.24 °/oo	NA	NA
1976	0/754		1/825 (1.21 °/oo)	3/1630 (1.84 °/oo)	NA
1977	7/550 (12.72)		16/695 (23.02)	15/1529 (9.81)	NA
1978 (I–VI)	5/298 (16.7)		7.321 (21.8)	10/716 (13.97)	NA

[a]NA = Not Available yet.

Therefore, a dramatic increase in BD in the whole area has been observed, suggesting that "something has happened". But several points are conflicting with the possibility of associating such a rise with the accident:

1. In 1977 the rate in Meda and Desio, which were less polluted, was twice as much as compared to more polluted Seveso and Cesano; again, in 1978 the highest rate was in the 7 cities where less than 2% of the population was involved in the polluted area.
2. A simultaneous increase of reported BD took place in several counties of Lombardy (Table VII): it has doubled in Mantova and Varese, tripled in Sondrio, and increased in four other counties, possibly due to the improvement in notification accuracy (according to the recommendations issued by the health authorities);
3. The distribution of BD by type shows no clustering and a great variety of BD compared to their total number (Table VIII);
4. The rates observed in 1977 and 1978 in the different groups of cities, up to 23.0 × 1000, are never higher than the ones usually observed in those western countries where notification systems work. This is confirmed also by a longitudinal study in Italy sponsored by the National Research Council: the rate on 2409 newborns in four Maternity Centers, 23.2 per 1000, is one of the same magnitude observed now in the Seveso area [5].

The BD rates for 1977, computed by pollution zone (Table IX), are 12.45 (non R) and 23.93 (zone R). No BD were observed in zone B (out of 66 births) or in zone A (out of 4 births).

It would be interesting to know whether the lack of BD in zones A and B can be attributed to the small number of births and whether the hypothesis of an increased rate is still compatible with such observed negative results.

Table VII. BD Rates (°/oo) in Lombardy and its Counties (Data for 1970–1975 [4]. Years 1976–1977, courtesy of Dr. V. Carreri, Assessorato alla Sanità, Regione Lombardia—unpublished data)

Counties	1970	1971	1972	1973	1974	1975	1976	1977
Brescia	4.01	4.75	4.33	4.91	4.40	3.62	3.70	3.14
Bergamo	3.08	3.37	2.41	1.97	3.44	5.02	5.14	3.62
Como	3.82	4.11	5.64	5.52	4.56	5.61	6.39	6.50
Cremona	7.21	6.54	4.17	6.20	2.09	5.60	6.44	6.93
Manitova	1.16	3.97	0.59	1.75	2.22	1.94	1.38	3.84
Milano	0.74	2.00	3.08	3.49	3.44	3.86	4.13	4.99
Pavia	6.03	3.73	N.A.	4.01	5.46	4.38	5.59	7.48
Sondrio	3.09	3.02	3.12	2.09	1.79	3.31	2.07	6.02
Varese	1.80	1.73	1.51	1.92	2.20	2.82	1.74	4.45
Lombardy	2.25	2.94	3.01	3.52	3.50	4.01	4.15	4.85
Italy	1.9	1.8	1.8	1.6				

Table VIII. BD Observed in the Polluted Area, 1976–1978 (I–VI)

A. Single Defect	ICD (1975)	No. in 1976	No. in 1977	No. in 1978 (I–VI)
Inguinal Hernia	550.9			1
Meningocele	741.9			1
Hydrocephalus	742.3		1	
Bulbus cordis anomaly	745		1	4
Interventricular septal defect	745.4		3	1
Stenosis of pulmonary valve	746.0		2	
Cleft palate	749.0			1
Stenosis small intestine	751.1	1		
Anomaly intestinal fixation	751.4	1		
Ectopic anus	751.5	1		
Hypospadia	752.6	2	2	1
Congenital anomaly of scrotus	752.9			1
Clubfoot	754.7		9	3
Syndactyly	755.1		2	3
Osteogenesis imperfecta	756.5		1	
Diaphragmatic hernia	756.6			1
Down's syndrome	758.0	2	2	1
Hamartoblastoma cordis	759.6		1	
Neuroblastoma	M/9500/3	1		
B. Multiple Defects				
Interventricular septal defect and ostium secundum	745.4			
type atrial septal defect	745.5		1	1
Interventricular septal defect and agenesis of lung	745.4			
	748.5		1	
Bulbus cordis anomaly and epispadia	745			
	752.6		1	
Meningocele and hydrocephalus	741.9			
	742.3		1	
Atresia of auditory canal and anomaly NOS of ear	744.0			
	744.3		1	
Clubfoot and syndactyly	754.7			
	755.1		1	
Exstrophy of urinary bladder and hypospadia	753.5			
	752.6		1	
Omphalocele and gastroschisis	553.1			
	756.7		1	
Anencephalus and omphalocele and spina bifida	740.0			
	553.1			
	756.1		1	
Clubfoot and clubhand omphalocele	754.7			
	754.8			
	553.1		1	
Bulbus cordis anomaly and Down's syndrome	745			
	758.0			1
Down's syndrome and polydactyly	758.0			
	755.0			1
Polydactyly and ectopic anus and cyst of urachus	755.0			
	751.5			
	753.7			1

Table IX. BD/Births by Pollution Zones and Year [1976 to 1978 (I–VI)]

Year (months)	Zone A	Zone B	Zone R	3 C.S.Z. (s̄ A;B;R)
1976 VII-XII	0/2	0/29	0.221	1/1280 (0.78%)
1977	0/4	0/66	9/376 (2.39%)	29/2328 (1.25%)
1978 (I-VI)	0/ ..	1/ ..	5/ ..	16/ ..

Table X. Probability of No Cases of Birth Defects in 1977 in Zone A and Zone B, Given Several Assumed Incidence Levels [6]

		Zone A	Zone B
1977	Observed BD	0	0
	No. births	4	66

Assumed BD Incidence Levels (x100)	Zone A		Zone B	
	No. Cases of BD Expected	Probability of No BD	No. Cases of BD Expected	Probability of No BD
2.4	0.096	0.907	1.6	0.201
5.0	0.2	0.815	3.3	0.034
10.0	0.4	0.660	6.6	0.001
20.0	0.8	0.410	13.2	4×10^{-7}
50.0	2.0	0.060		

In Table X, probabilities have been calculated that there were no cases of BD among newborns of Zones A and B given several assumed BD incidence levels.

Little can be concluded for zone A: probability that the incidence has risen to 50% is only 6%, but the hypothesis cannot be rejected that the incidence has reached 5, 10 or even 20% because the probability of no BD out of 4 births is still, respectively, 81.5, 66 and 41%. On the other hand, there is still a 20% probability that no BD are observed in Zone B out of 66 births, should the incidence be equal to the one in zone R (namely 2.4%); but there is only a 3.4% probability should the incidence have risen up to 5%.

Therefore, in spite of the low number of births in zone B, among which no BD have taken place, it can be concluded that the incidence in zone B is not higher than in zone R. Nothing can be said for zone A, with only 4 births, except that should an increase have taken place, this would have not been higher than 20%!

One observation to conclude this part. It has been noted that the lack of increase of BD in the area could be due to the fact that several women, fearing the fate, underwent a kind of "therapeutic" abortion; nothing can be said for the ones who aborted secretly or abroad. But for the 30 official abortions, which were all studied by Prof. Gropp of ETS, 4 were women of zone A, 3 of them were exposed as long as 14 days before evacuation. No malformations were found in their embryos [7]. Therefore, in spite of some incertitudes intrinsic to embryological diagnosis of BD, it can be concluded that at least 8 pregnancies in zone A did not involve any BD.

SPONTANEOUS ABORTIONS

To evaluate the possible effects of TCDD pollution on the frequency of spontaneous abortion (SA), two lines of investigations are being considered: the first one consists in the registration of the SA reported by physicians to the county medical officer; the second one integrates the notifications with a special search on admission/discharge hospital forms. The first study allows comparisons with regional and national statistics, when available, while the second one is an ad hoc project.

In Table IX SA rates per 100 pregnancies per year are shown. Rates fluctuated from 1973 to 1977. In 1977 rates are always higher than 1976; nevertheless, they are lower or rather similar to the ones of previous years. It is not easy to explain this rise, which could be affected by improved physicians' care in the reporting. In fact, such a rise is common to all the 11 cities and is not a very large one.

Figure 2 compares polluted area rates with Milano and Lombardy rates. The trend of polluted area rates is very similar to Milano and Lombardy rates, which are always higher, until 1976. In 1977 there is a rise in the polluted area, but even the improvement of SA notifications did not permit the rate to reach the level of Milano and Lombardy. About the second line of investigation, the aim is to refer SA to total conceptions of the same period.

SA of a given trimester, collected by the county medical officer, and the hospital forms constitute the numerator. The denominator includes the same SA plus births and stillbirths with the same time of conception.

Table XI. Spontaneous Abortion Rates (x 100 pregnancies) from 1973 to 1977

Cities / Years	1973	1974	1975	1976	1977
Cesano M. Seveso	9.55	9.76	10.04	10.13	10.19
Desio Meda	12.34	10.57	10.81	9.01	12.31
Total, 4 cities	10.91	10.15	10.45	9.54	11.38
Other, 7 cities	10.88	10.90	9.64	10.11	11.81
TOTAL, 11 cities	10.89	10.55	10.01	9.74	11.63

$$\text{Pregnancy loss rate (PLR)} = \frac{\text{Abortions (Ja + F + M)}}{\text{Births + Stillbirths (Ju + A + S) + Abortions (Ja + F + M)}} \times 100$$

Data on births and stillbirths were collected from Census Bureau and Hospital forms. Induced abortions were excluded from numerator and denominator.

Such a system of collecting data became possible only after July 10, 1976, when the ICMESA accident happened: in fact, after this date hospital forms related to people living in the polluted area were filed separately from all the others. Data were processed by hand because in the regional computer, working then for administrative rather than epidemiological purposes, data were fed into it in the "without names" format.

Table XII shows the pregnancy loss rate (PLR) by trimester from July 1976 to December 1977. In the cities with more than 50% of the inhabitants in the polluted area, the rates show a rise in the fourth trimester of 1976; then they decrease in the following trimesters. In the cities with about 20% of inhabitants in the polluted area, the rates increased until the first trimester 1977; then they decreased. A similar trend is visible in the other seven cities.

Such a trend is very hard to understand because data accumulated before the accident are not available. It will be difficult to obtain these data because of hospital form information system criteria. Moreover, should difficulties be overcome, the earliest comparison period would go back no farther than January 1976, the date routine hospital form collection was begun in Lombardy.

A more satisfactory definition of exposure levels is now in progress. Table XIII shows PLR by pollution area (Zone B, R and other zones):

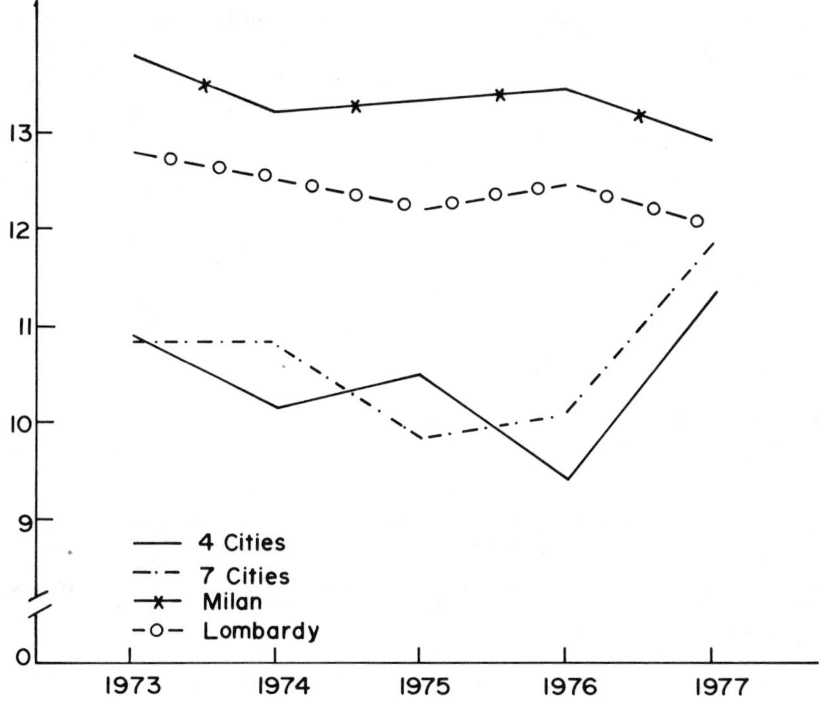

Figure 2. Abortion rate (x 100) pregnancies) from 1973 to 1977).

Zone A data were excluded because of the small number of pregnancies. PLR for Zone B are always higher than in other zones. Only in the third trimester of 1977 did rates show statistically significant differences ($P<0.005$). Nevertheless, such a difference did not exist among various rates in Zone B.

CRUDE BIRTH AND DEATH RATES

Data on birth rates (BR) and deaths (DR) were provided by the Census Bureau. BR and DR were calculated per year and per thousand inhabitants. Table XIV shows birth rates from 1973 to 1977: they are provided for three groups of cities differing from each other by the amount of inhabitants in polluted areas, as shown previously. BR generally decreased from 1973 to 1977.

Table XII. Pregnancy Loss Rates (PLR) by Trimester from July 1976 to December 1977

Cities		Jul-Sept 1976	Oct-Dec 1976	Jan-Mar 1977	Apr-Jun 1977	Jul-Sept 1977	Oct-Dec 1977
Cesano M. Seveso	A[a]	27	30	21	21	24	25
	P[b]	164	141	188	171	186	186
	R[c]	16.5	21.3	11.2	12.3	12.9	13.4
Desio Meda	A	18	26	33	27	25	32
	P	200	187	208	197	198	237
	R	9.0	13.9	15.9	13.7	12.6	13.5
Other 7 cities	A	50	58	84	59	44	66
	P	427	414	462	418	423	443
	R	11.7	14.0	18.2	14.1	10.4	14.8

[a]A = Abortions.
[b]P = Pregnancies.
[c]R = PLR.

Table XIII. Pregnancy Loss Rates (PLR) by Pollution Zones, by Trimester from July 1976 to December 1977

Cities		Jul-Sept 1976	Oct-Dec 1976	Jan-Mar 1977	Apr-Jun 1977	Jul-Sept 1977	Oct-Dec 1977
Zone B	a[a]	3	4	5	8	10	4
	P[b]	27	18	29	28	32	29
	R[R]	11.1	22.2	17.2	28.5	31.2	13.7
Zone R	A	19	17	15	17	16	20
	P	138	104	118	135	140	144
	R	13.7	16.3	12.7	12.5	11.4	13.8
3 C.S.Z. $\bar{s}$ A,B,R	A	74	94	119	81	67	99
	P	670	632	713	621	634	691
	R	11.0	14.8	16.6	13.0	10.5	14.3

[a]A = Abortions.
[b]P = Pregnancies.
[c]R = PLR.

In Figure 3 such a decrease is quite evident until 1976, mainly in the two cities in which more than 50% of inhabitants were in polluted areas. There is definite evidence of a reduction of BR in 1977 in the four most polluted cities, probably because of birth control, which authorities advised for about one year after the July 1976 accident.

Table XIV. Birth Rates (x 1000) from 1973 to 1977

Cities / Years	1973	1974	1975	1976	1977
Cesano M. Seveso	18.78	17.51	14.86	13.62	10.94
Desio Meda	17.44	16.63	16.54	15.77	13.20
Total, 4 cities	18.21	17.07	15.71	15.36	12.10
Other, 7 cities	17.68	17.15	16.68	14.14	13.22
TOTAL, 11 cities	17.93	17.11	16.22	14.72	12.66

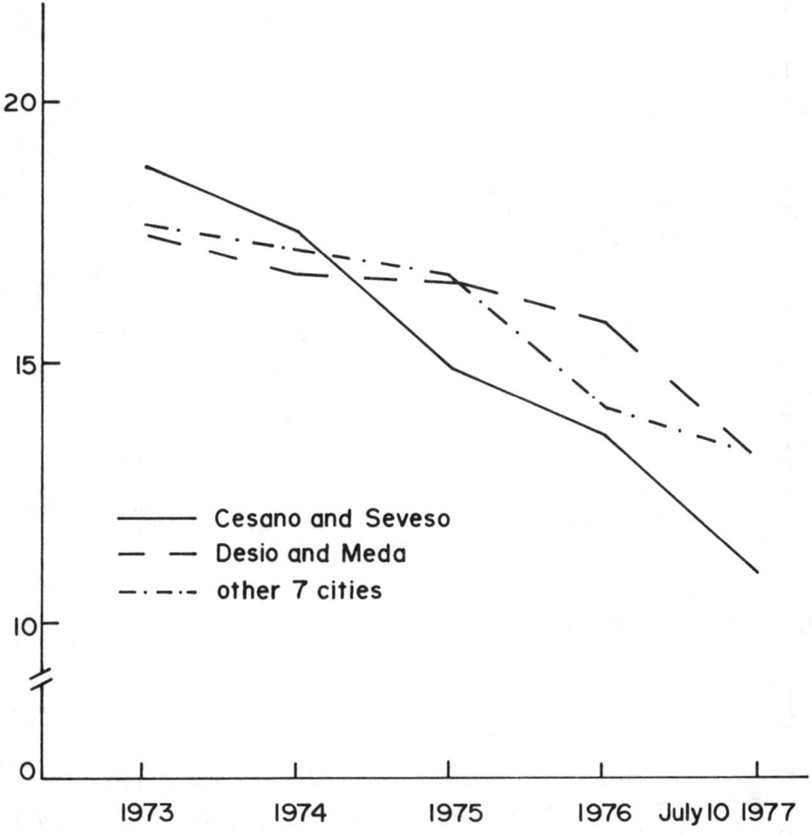

Figure 3. Birth rates (x 1000) from 1973 to 1977.

Figure 4 shows BR in four cities and seven cities, and in Lombardy and Italy. In the polluted area, BR were generally higher than in Italy and in Lombardy. The most interesting phenomenon was seen in 1977, when BR for the more polluted area decreased more sharply in comparison with previous years.

About mortality, we can provide only crude DR. It has been impossible so far to calculate specific DR per cause of death, age and sex because of the lack of information on these factors. Nevertheless, we are studying a specific research program to obtain DR per age and sex, while rates per cause of death will be available only from now on.

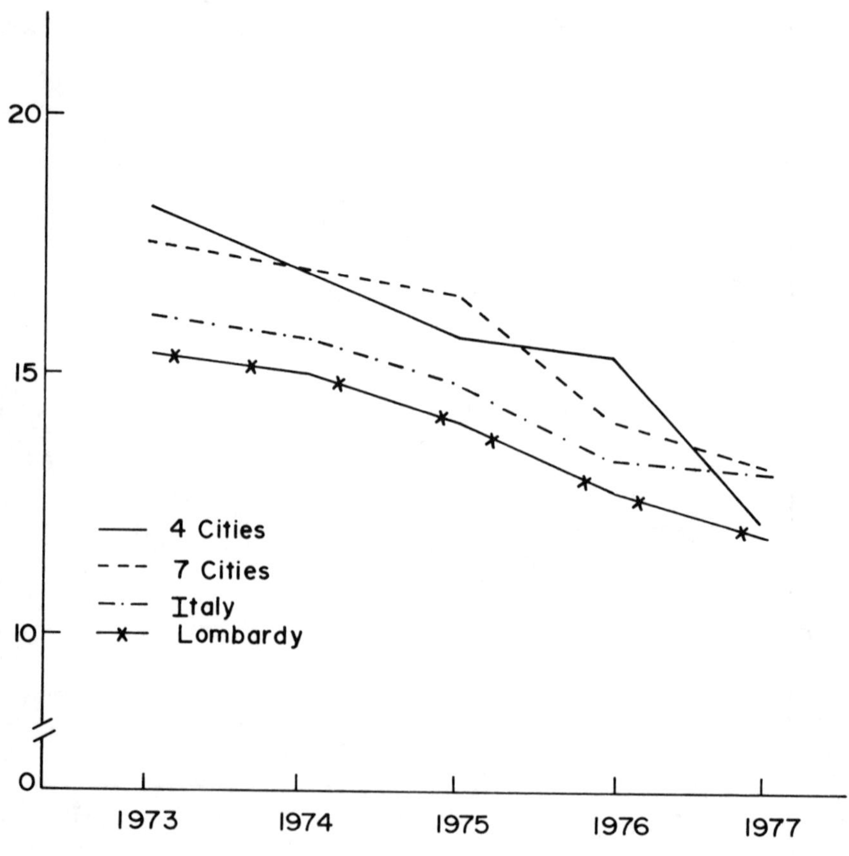

Figure 4. Birth rates (x 1000) from 1973 to 1977.

Table XV shows crude DR per year for the three groups of cities. We can observe differences in the DR trend at Cesano and Seveso, more than 50% of whose inhabitants were in the polluted area. Rates increased at Cesano until 1975; on the other hand they decreased in 1976–77. At Seveso the mortality rate (MR) decreased until 1975 and rose in 1976; then it decreased again in 1977. In the cities in which 20% of the inhabitants were in the polluted area (Desio and Meda), rates are irregular, with few variations. In the other seven cities, crude DR are stable, except for a rise in 1976.

Figure 5 shows the same data for the four cities and for the seven cities; an increased DR was registered in 1976. Because of this rise, we suspected a possible association with the TCDD pollution. Nevertheless, a six-month-period disaggregation of DR in 1976 (Table XVI) shows that rates were generally higher in the first half year before the July 10 accident. This phenomenon occurred in all 11 cities, particularly in Seveso. The same phenomenon also appeared in 1977, except in Seveso where the rate was higher in the second half year.

Figure 6 shows crude DR for the 4 cities and the seven cities vs Italy and Lombardy. DR in Italy and Lombardy were higher than in the polluted area. Moreover, Lombardy shows the same slight rise in crude MR observed in the polluted area.

To conclude, birth rates in the area show a steeper decrease in 1977, probably due to birth control, and crude DR do not seem to have been affected at yet by the TCDD pollution.

Table XV. Crude Mortality Rates (x 1000) from 1973 to 1977

Cities / Years	1973	1974	1975	1976	1977
Cesano M.	6.28	6.65	7.13	7.07	6.51
Seveso	9.18	7.50	6.83	9.55	7.72
Desio	8.92	8.27	8.86	9.05	7.69
Meda	8.48	7.72	8.33	8.25	8.71
Total, 4 cities	8.00	7.51	7.86	8.33	7.51
Other, 7 cities	7.91	7.94	7.80	8.36	7.69
TOTAL, 11 cities	7.95	7.73	7.83	8.35	7.61

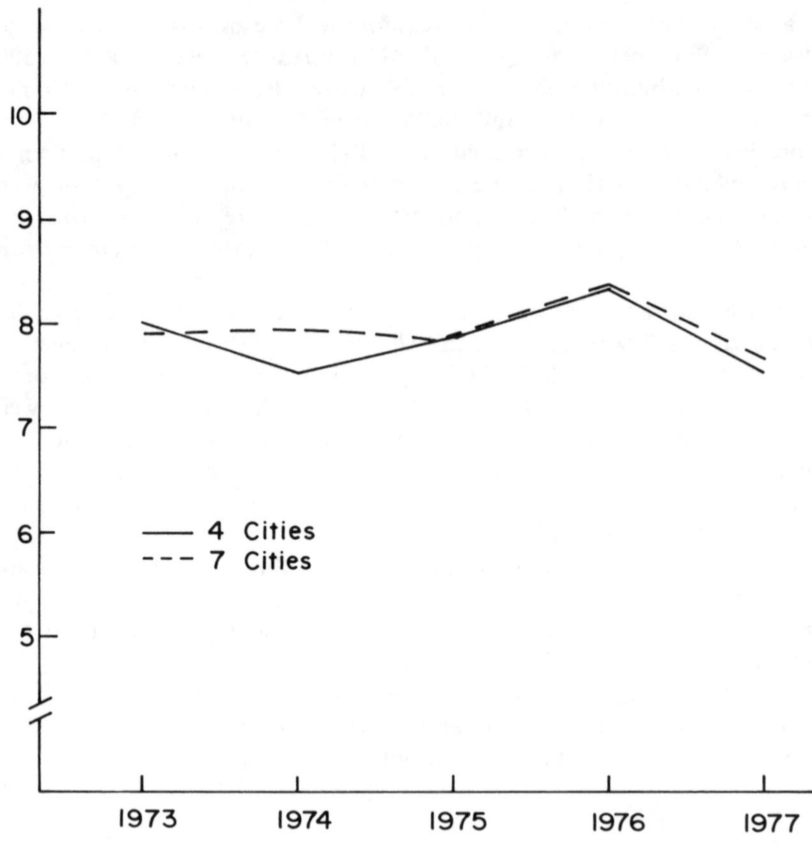

Figure 5. Crude mortality rates (x 1000) from 1973 to 1977.

Table XVI. Crude Mortality Rates (x 1000) by Six-Month Periods in 1976 and 1977

Years	1976			1977		
Cities	**Jan-Jun**	**Jul-Dec**	**Jan-Dec**	**Jan-Jun**	**Jul-Dec**	**Jan-Dec**
Cesano M.	3.64	3.43	7.07	3.57	2.94	6.51
Seveso	5.66	3.89	9.55	3.65	4.06	7.72
Desio	5.16	3.88	9.05	4.14	3.54	7.69
Meda	4.38	3.86	8.25	4.43	4.28	8.71
Total, 4 cities	4.50	3.73	8.33	3.93	3.58	7.51
Other, 7 cities	4.45	3.91	8.36	3.88	3.81	7.69
TOTAL, 11 cities	4.52	3.83	8.35	3.91	3.70	7.61

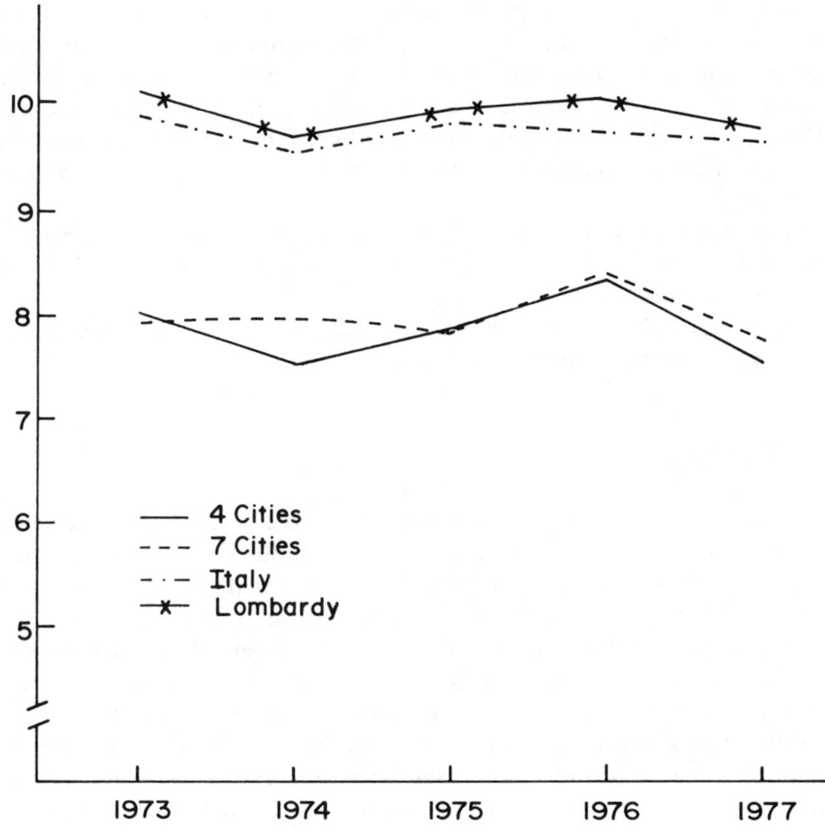

Figure 6. Crude mortality rates (x 1000) from 1973 to 1977.

SPECIFIC RESEARCH PROGRAMS

Specific research programs dealing with immunology, neurology and cytogonetics are being developed by ad hoc teams. Only progress reports have been issued and they are quoted below:

Immunology. 45 children previously living in zone A, 21 of them with chloracne of the "early" type, together with as many age-matched unexposed controls, were submitted to a battery of tests to investigate their immune response. This was started in September 1976, and tests were repeated three times at four-month intervals. According to a statistical evaluation of the results, no significant differences between exposed children and controls could be demonstrated [3].

Neurology. Instrumental examination of groups of persons have shown higher rates of subclinical neurological damages in zone A than in zones B and R, namely reduced nerve conduction velocity, which could not be related to any known cause. This was confirmed also during another screening in 1978. No association exists between the occurrence of the skin lesions and the neurological findings. A careful evaluation of these findings is in progress [8].

Cytogenetic tests. They were conducted on peripheral blood lymphocytes stimulated by PHA and obtained from 125 people of Zone A, 59 of ICMESA workers and 69 controls.

Neurological, immunological and cytogenetic studies are still in progress to determine whether any alteration will appear in the longer term.

CONCLUSION

The contents of the preceding pages represent a first report on selected topics prepared by the working team appointed by the regional government of Lombardy. The joint efforts, started at the end of July 1976 and based on a plan prepared to cope not only with the need for epidemiological knowledge but also with the urgent demand from the population involved, immediately have faced great difficulties.

Obstacles were many and great, beginning with the availability of poor quality demographic data and with the lack of local public health organization; progressing with the sudden political struggle for and against the abortion policy to be applied to local women and with a lot of nonsense solutions proposed by many so-called scientists, at home and abroad to terminate the pollution; and ending up with the progressively spreading resistance of the population to cooperate in the surveillance plan.

Therefore, not all data are available that could help in preparing a final picture of all the health events that followed the July 10, 1976 accident. Some of them are gone forever, others could be collected and accumulated but their elaboration was delayed because of lack of operators and the call for more urgent activities. Systematic analysis and elaboration of available data could only start at the beginning of 1978, when the working team in Seveso was finished.

Interpretation of data presented here does not seem to indicate that any major event has happened yet in the field of birth defects, abortions, births and deaths (crude data). Whether because the TCDD exposure was per se limited or because of the efficacy of the health measures immediately enforced, including the evacuation of zone A residents, is hard to say. As far as the chloracne is concerned, almost all serious ("early") cases were detected in people of zone A (or living there during the accident); in spite of the general recommendation to call a physician, no cases were reported in zone B and only one in zone R, which had a history of contact with zone A. "Late" cases (by far the least serious) were mostly detected through screening, but their distribution was not confined to zones A or B; it covered zones R and non(A,B,R) also, with higher prevalence in zones R of Seveso and Meda (closest to zone A) than in zone B. Such lack of overlapping of the chloracne map with the map of TCDD in soil is now under consideration, employing the exposure questionnaires administered to the chloracne patients. Apart from the evidence that the Seveso and Meda residents of zone R, during the days after the accident and before its evacuation, had easy access to, and contact with, the area classified later as A, a possibility exists that the July 10th cloud has intersected the atmosphere of a larger area around the factory than is now suspected according to the detection of TCDD deposited on the ground. This will only be clarified after the ICMESA reactor, now under sequestration by the Court, is studied for its contents.

Additional studies will be needed to clarify the significance of atrophodermy and other lesions resembling healed chloracne, which were diagnosed and classified as dating back before July 10, 1976.

Additional data will also be needed to interpret the functional signs of neurological involvement found with higher prevalence in zone A residents than in other groups and to explain their clinical significance and the apparent lack of correlation with chloracne lesions.

Finally, although the immunological and cytogenetic studies carried out on subjects both with and without chloracne have not shown any significant alteration so far, there is a wide agreement as to the opportunity of their continuation because it cannot be excluded that some abnormality could appear in the longer term.

ACKNOWLEDGMENTS

The Authors acknowledge Drs. A. Andreani, G. Beltrami and S. Sicurello for their cooperation of data processing; and Mrs. S. Blanco, F. Formigaro, M. Mauri and A. M. Rosa for their clerical work.

REFERENCES

1. *Corriere della Sera* (September 1, 1978).
2. Fara, G. M. "Preliminary Data on the Accident in Seveso," in *Round Table on*

the Seveso Accident. (M. A. Klingberg, Chairman), Proc. 5th Conference of the European Teratology Society (in press).
3. Sirchia, G. "Relazione riassuntiva sull'attività di ricerca clinica nel settore immunologico," Milano (May 31, 1978).
4. Carreri V., and A. Buratta. "Andamento epidemiologico della rosolia negli anni 1971–1975. Confronto con i casi di aborto e di nati malformati in Lombardia," *Ann. Sclavo,* 18:714–19 (1976).
5. Fara, G. M., and E. Marubine. "Monitoring Birth Defects: an Italian Project," *Proc. 3rd Conference European Teratology Society, Abst. Teratol.*, 10:309 (1974).
6. Modlin, J. F., K. Herrmann, A. P. Brandling-Bennet, D. L. Eddins and G. F. Hayden. "Risk of Congenital Abnormality after Inadvertent Rubella Vaccination of Pregnant Women," *New England J. Med.* 294:972–74 (1976).
7. Rehder, H., L. Sanchioni, F. Cefis, and A. Gropp. "Pathologisch-embryologische Untersuchungen an Abortusfällen im Zusammenhang mit dem Seveso-Unglück," *Schweia. Med. Wochenschr.* 108:1617–1625 (1978).
8. Boeri, R., G. Filippini, M. Massetto, B. Bordo, P. Crenna and A. Zecchini. "Preliminary Results of a Neurological Investigation of the Population Exposed to TCDD in the Seveso Region," Incontro Italo-Polacco, Varenna (Como) 17–18 June 1978 (1978).
9. Morganti, G. "Relazione conclusiva sull'attività di ricerca clinica nel settore genetico," Milano, (July 31, 1978).

EPILOGUE

James T. Murphy
Principal Environmental Engineer
County of Bergen
Hackensack, New Jersey

An epilogue is defined as "A concluding section that rounds out the design of a literary work." The dialogue set forth at the conference stimulated the mind and certainly focused on the magnitude and long-term relationships of hazardous waste management. Regrettably, however, the conference was more of a *prologue*: "An introductory or preceding event or development." The substance of most papers indicated that systems are available to meet the problems, respond to the condition and manage the solution.

However, the "regulations" dominate any solution. In my opinion, a real world enterprise with a hazardous waste problem cannot hope to go "one-on-one" with a "regulation." The substance of the regulation is subject to the interpretation of the enforcing agency and the personal opinions and prejudices of the individual reviewing the matter. Ambiguity of language, open-ended, vague interconnected relationships make it almost impossible to arrive at a sound engineering or chemical solution.

A fundamental principle that has stood the test of time is "Matter can neither be created nor destroyed." This becomes the ultimate "Catch 22" in the regulatory system. The second half of the principle, "We can only change its form," dictates the solution. Given a change of form, we immediately superimpose a new set of conditions that must comply with other regulations.

Obviously, the many processes and manufacturing lines that are necessary to sustain our way of life will not come to a screeching halt because of regulations. Technology and American knowhow will find a way to legally pursue their endeavors and make the regulations work for them.

INDEX